# RIGID AND SEMI-RIGID PLASTIC CONTAINERS

J. H. BRISTON

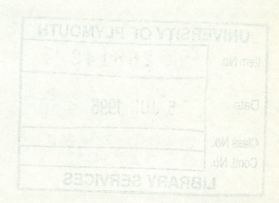

**Longman Scientific and Technical**
Longman Group UK Limited
Longman House, Burnt Mill, Harlow,
Essex, CM20 2JE, England
*and Associated Companies throughout the world.*

*Copublished in the United States with
John Wiley & Sons, Inc., 605 Third Avenue, New York, NY 10158*

© Longman Group Limited 1994

First published 1994

**British Library Cataloguing in Publication Data**
A catalogue record for this book is available from the British Library

ISBN 0 582 01491 3

**Library of Congress Cataloging-in-Publication Data**
A catalog entry for this title is available from the Library of Congress

ISBN 0 470 23423 7

Set by 16JJ in 10/12½pt Times

Printed and bound in Great Britain
by Bookcraft (Bath) Ltd

# Contents

# CHAPTER 1

# Introduction

This book deals with the manufacture and use of rigid and semi-rigid plastic containers together with an account of the materials from which they are made. Mention is also made of certain multi-material containers where the plastic component plays a significant role. The philosophy of using two or more materials in combination is that quite often no single material will supply the properties needed to produce a satisfactory package to meet the increasingly complex demands of modern products, marketing and packaging machinery. It may also, sometimes, be possible to achieve appreciable cost savings.

The title, '*Rigid and Semi-rigid Plastic Containers*', was chosen after much thought and discussion. The intention was to cover such containers as bottles, pots, tubs and cups, trays and tote boxes. A shorter and less clumsy title would have been simply, 'Rigid Plastic Containers', but comments received showed that some people would have perceived the book to be one covering high-performance glass-reinforced polyester tanks and similar containers so the addition of '... and Semi-rigid ...' was made.

Whatever the description given to the range of containers under consideration, there is no doubt about the importance of the markets involved. If we consider, for instance, the example of the latest bottle-blowing material, poly(ethylene terephthalate) or PET, an estimated 6 billion ($6 \times 10^9$) bottles were produced world-wide in 1986, and predictions have been made of 27.6 billion PET containers per annum by 1995. All this stems from the introduction of the 2-litre soft drinks bottle in 1976. When one considers the equally large numbers of containers manufactured from polyethylene, PVC, polystyrene, polypropylene, etc., the importance of the subject becomes clear.

An even clearer indication of the importance of plastic containers today is the fact that whereas plastics were once considered to be substitute materials (and cheap and nasty ones at that!), they are now often the first choice when a new product has to be launched. It must be remembered, too, that this change of attitude has taken place during a time when the prices of plastics are no longer dropping steadily *vis-à-vis* other packaging materials. Today, any new use for plastics has to justify itself on all-round performance/cost characteristics.

One of the advantages of plastics in general is their design versatility. Not only is there a wide range of plastics available (with a correspondingly wide range of properties), but there are many processes by which plastics can be converted into useful package forms. Thus, there is injection moulding which can be used to produce articles ranging from a small aerosol valve button to a tote box or a crate. Extrusion and injection blow moulding are equally versatile, producing containers as small as a 5 ml vial or as large as a 210 litre barrel or drum. In addition, it should be noted that both wide-and narrow-neck containers can be produced by blow moulding. Plastics can also be extruded into sheet and this can be further fashioned into cups, tubs, pots, shallow trays and box inserts (such as are used in the packaging of chocolates or biscuits).

Part 1 of this book is devoted to the properties of the plastics used in packaging from the point of view of aiding material selection, once the performance parameters are known.

Part 2 deals with the various methods by which plastics are converted into the various package forms, together with some notes on the choice of method. Ancillary processes, such as printing and decorating, are also described.

Part 3 brings the two strands (materials and conversion) together and outlines the various applications for rigid and semi-rigid plastic containers, under headings such as food and drink, pharmaceuticals, cosmetics and toiletries, etc.

Part 4 deals with the ways in which the various properties of plastics (already described) are determined and assesses their significance to the converter or end user.

Finally, there are two appendices dealing with (a) some of the additives to be found in plastic materials and (b) the subject of closures.

One subject that is not dealt with in the following chapters is that of recycling and reuse of plastic containers. This is not because there is little said or written about the subject. On the contrary, it has been obsessively examined and reported on by governments, environmental groups and by industry. The issue is an emotive and a political one and is best studied by reading widely. In addition, there is a great deal of current and potential legislation and the picture is constantly changing. For example, one item still under discussion is the question of the incineration of plastics. This is seen as wholly undesirable by environmental organisations while recycling is presented as the cure for all ills. However, there are some criteria for the *successful* recycling of post-consumer waste, the most important being that the recycling operation should provide a net environmental benefit overall. If any particular recycling operation results in an *overall energy loss* then that particular operation is environmentally damaging. Incineration, on the other hand, can be made to provide useful energy. A recent report by the Royal Commission on Environmental Pollution contains the following statement: 'We are satisfied that the incineration of waste in suitably located plants, designed to meet new standards, represents an environmentally acceptable form of waste disposal.'

An example of what can be achieved is given by a waste-processing plant operating in Italy (on the shores of Lake Maggiore). A combination of gasification and incineration at temperatures of up to 2000 °C, converts municipal waste into a gas that can be used for the plant's own energy needs or for electricity. The residue consists of recyclable metals and minerals. It is claimed that 99.5% of the waste is regained, either as energy or as reusable raw materials. The process is also said to be as near emission-free as possible and to be cheap to run.

Common sense would seem to dictate that the best use is made of any available technology in order to minimise harmful environmental impacts due to the use of packaging but common sense is in noticeably short supply where environmental issues are involved.

A further area of concern is that of food contact safety and those readers interested in this subject should consult the two chapters written by Dr L. L. Katan in the author's book, *Plastics Films*, 3rd edition, 1989, Longmans, London. It should be noted that here, again, the subject is in constant flux, both in the area of legislation and in that of research.

*PART 1*

---

MATERIAL SELECTION

# CHAPTER 2

# Polyolefins

The polyolefins are a very important class of materials, the main members of which are polyethylene, polypropylene and some copolymers.

## 2.1 Polyethylene

Polyethylene is a thermoplastic polymer based on the polymerisation of ethylene ($CH_2 . CH_2$) and is available in a range of densities. Low-density polyethylene (LDPE) is usually taken to cover materials with a density between 0.915 and 0.939 g/cm$^3$ while high-density polyethylene refers to a density of 0.940 g/cm$^3$ and above.

For any particular density of polyethylene there is a range of different molecular weights. These materials vary, particularly in processability and in mechanical properties and it is, therefore, essential for the converter to have some measure of the molecular weight of the particular polyethylene in which he is interested. A simple method that depends on measurement of the melt viscosity of the polymer was developed by Imperial Chemical Industries plc. This has since been modified and accepted as a British Standard (BS 2782: Part 1) while a similar method has been adopted as an ASTM procedure (ASTM D 1238). A schematic diagram of a melt indexer is shown in Fig. 2.1.

The apparatus is essentially an extrusion plastometer with a standard orifice. The weight, in grams, of material extruded in 10 min under a constant dead-weight load, at a temperature of 190 °C, is called the melt flow index (MFI). The pressure is quite low (303 kPa) so that although the results are very useful as a method of grading, there are limitations due to the fact that the flow of molten polymer is non-Newtonian. Because of this it is impossible to forecast the flow behaviour under the high shear conditions that are obtained in commercial extruders and injection moulding machines from the results of tests using the very much lower shear forces in the melt indexer. Other equipment is available, however,

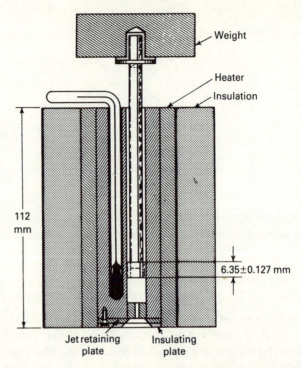

**Figure 2.1**  Apparatus for determining melt flow index. (Reproduced from Briston, *Plastics Films*, 3rd edition, 1989, Longman, London.)

that can simulate the conditions existing in an extruder or injection
moulding machine.

A low melt index corresponds to a high melt viscosity. Since melt
viscosity is directly related to the molecular weight of the polymer, a low
MFI corresponds to a high molecular weight. A strict relationship applies
only to polymers of similar density because density also affects melt
viscosity. For polyethylene of a higher density, a particular MFI will
correspond to a slightly smaller molecular weight. The MFI is effectively
used as a method of grading polyethylenes and, together with density, is
a good guide (though not a complete one) to the type of polyethylene
under consideration.

A further parameter that aids characterisation is the molecular weight
distribution (MWD). The molecular weight of any given sample of resin
is only an average figure and the range can sometimes be quite large. In
general, a wide MWD denotes lower impact strength, lower melt strength,
better processability and higher resistance to environmental stress cracking (see later on in this chapter).

The density of a polyethylene bottle or other moulded article is controlled by the density of the polymer used in the moulding; the cooling rate is also important. The faster the cooling rate, the less time is available for the polymer chains to crystallise and the lower the density of the final article. Density affects such properties as rigidity, stiffness, softening point, water vapour transmission rate and environmental stress crack resistance.

*Manufacturing process*

Before looking at the properties of polyethylene in greater detail, it will be helpful to look briefly at the manufacturing processes.

Low-density polyethylene (LDPE) is made by the high-pressure polymerisation of ethylene. Either tubular or autoclave reactors are used, with pressures of the order of 1000 to 3000 atmospheres and temperatures of 150 to 250 °C. Small amounts of an initiator (usually oxygen or a peroxide) are added to start the process. Other monomers such as vinyl acetate or ethyl acrylate can be added to produce the materials ethylene/vinyl acetate copolymer (EVA) and ethylene/ethyl acrylate copolymer (EEA). Molecular weight, molecular weight distribution and density are controlled by reaction temperature, pressure of ethylene and the concentration of chain transfer agents.

There are slight differences in the products of the tubular and autoclave processes. The tubular reactor produces LDPE with a large number of long chain branches (over six carbon atoms in length) though these are still relatively short compared with the autoclave reactor products. The latter have a smaller number of long chain branches but they are of very much longer length. In general, the tubular process gives products of greater clarity and better processing characteristics, while the autoclave reactor gives products that are useful in high-speed extrusion coating and in film applications requiring toughness.

High-density polyethylene (HDPE) can be made by a number of different processes. The original processes (Ziegler and Phillips Petroleum) used relatively low pressures and fairly high temperatures and the product was obtained as a slurry or in solution. During the late 1960s, Union Carbide Corporation introduced a gas-phase process using pressures of around 1960 kN (285 lb/in$^2$) and temperatures within the range 85–100 °C.

HDPE homopolymer has a largely linear structure with a small number of short chain branches. Such a structure allows close packing of the molecules and hence a large amount of ordered, tightly packed crystalline regions are formed during cooling to the solid state. These crystalline regions form as the long polyethylene molecules fold on themselves in a parallel arrangement to form lamellae. Crystallisation spreads as other molecules align themselves into position and also start to fold. Irregularities in molecular structure cause this crystalline growth to occur in more than one direction with the consequent formation of larger structures known as spherulites. The density of the largely linear HDPE is around 0.96 g/cm$^3$. Controlled short chain branching can be introduced by the incorporation of small amounts of other olefins as comonomers. The use of butene-1, for example, gives an ethylene/butene copolymer with ethyl (C$_2$H$_5$) branches. Depending on the amounts of comonomer added, the density is reduced to 0.94–0.95 g/cm$^3$ with a consequent reduction in melting point, stiffness, and barrier properties.

### 2.1.1 Physical properties

LDPE is a tough, slightly translucent, waxy solid. Increasing the density has the effect of raising the softening point, increasing the rigidity and surface hardness and decreasing the impact strength. The higher density materials are also less waxy in appearance and feel. One of the most significant differences between low- and high-density polyethylene is that of the softening point, since one is below the boiling point of water while the other is above. Articles manufactured from the high-density material can thus be steam sterilised, thereby opening up many applications that cannot be served by LDPE.

The increased rigidity of HDPE means that equivalent rigidity can be attained with lower wall thickness and consequent savings in raw material costs. In many cases now, materials having densities intermediate between those of low- and high-density polyethylene are used.

Typical physical properties of LDPE and HDPE are shown in Table 2.1 but it must be remembered that many of these can be affected by molecular weight and molecular weight distribution. These possible effects are summarised in Table 2.2.

*Table 2.1*  Typical physical properties of polyethylene.

| Property | Test method | Units | Value Low density polyethylene | Value High density polyethylene |
|---|---|---|---|---|
| Density | | g/cm$^3$ | 0.915–0.94 | 0.941–0.965 |
| Tensile strength | BS 2782: Part 3 | psi (M N/m$^2$) | 1015–2320 (7–16) | 3180–5505 (22–38) |
| Elongation | BS 2782: Part 3 | % | 100–600 | 50–300 |
| Impact strength (Izod) | BS 2782: Part 3 | ft lb/in. (J/m) | no break | 2–12 (107–642) |
| Softening point (Vicat) | BS 2782: Part 1 | °C | 85–87 | 127 |

*Table 2.2*    Effect on properties of density, molecular weight and molecular weight distribution (MWD).

| Property | Effect of density increase | Effect of molecular weight increase | Effect of MWD broadening |
|---|---|---|---|
| Melting/softening point | Increase | Slight increase | Slight decrease |
| Tensile strength | Increase | Increase | Decrease |
| Impact strength | Decrease | Increase | Decrease |
| Environmental stress crack resistance | Decrease | Increase | |

### 2.1.2 Chemical properties

In general, the polyethylenes are resistant to most chemicals except oxidising acids, free halogens and certain ketones. At room temperature, polyethylene is insoluble in all solvents. At elevated temperatures the solubility of polyethylene in hydrocarbons and chlorinated hydrocarbons increases quite sharply. The solubility in any particular solvent is dependent on the density of the polymer, the high-density material having the lower solubility. Although insoluble at room temperature, polyethylenes slowly absorb hydrocarbons and halogenated hydrocarbons with subsequent swelling. Here again, the effect is less for the high-density materials. Table 2.3 shows the resistance of LDPE and HDPE to certain common chemicals.

Owing to the paraffinic nature of the polyethylene molecule, it is extremely resistant to water; the water absorption at room temperature being so low as to be negligible. In this respect, there is very little difference between the low- and high-density polymers. The water vapour permeability is also low but here HDPE is a better barrier than is LDPE. Water vapour permeability is closely affected by crystallinity and this is primarily a matter of molecular structure, although it can be modified by extrusion conditions. The higher the crystallinity, the lower the permeability, although the relation is not linear. Permeabilities of LDPE and HDPE to water vapour and certain gases are shown in Table 2.4.

The resistance of LDPE to oils and greases is rather poor but, again, improvement is noted with increases in molecular weight and density.

Polyethylene is susceptible to oxidation, particularly in the presence of UV light. Prolonged exposure to heat also accelerates oxidation with consequent deterioration of electrical and mechanical properties and appearance. The addition of antioxidants (such as substituted phenols, aromatic amines, etc.) minimises oxidation due to high temperatures but photo-oxidation is best combated by the incorporation of about 2% of carbon black which screens out the UV light. For outdoor use in tropical countries, further protection is obtained by the addition of 0.1–0.2% of

*Table 2.3*  Effects of various chemicals on polyethylene (at room temperature).

| Chemical | % Weight increase (30 days immersion) | | Remarks | |
| | LDPE | HDPE | LDPE | HDPE |
| --- | --- | --- | --- | --- |
| Acetic acid (glacial) | 0.85 | 0.75 | Some attack | Some attack |
| Acetone | 1.25 | 0.65 | Some loss of strength | Some loss of strength |
| Alcohol (ethyl) | 0.20 | 0.15 | Some loss of strength | Some loss of strength |
| Ammonium hydroxide | Negligible | Negligible | Unaffected | Unaffected |
| Carbon tetrachloride | 36.0 | 14.5 | Swelling and loss of strength | Swelling and loss of strength |
| Ethyl acetate | 3.10 | 2.30 | Slight attack | Slight attack |
| Motor oil | 3.40 | 0.70 | Slight attack | Satisfactory |
| Nitric acid (conc.) | 0.35 | 0.30 | Some attack (yellowing) | Some attack (yellowing) |
| Petroleum | 13.0 | 6.0 | Swelling and loss of strength | Slight swelling and of strength |
| Sulphuric acid (conc.) | Negligible | Negligible | Some attack | Some attack |

*Table 2.4*  Permeability of polyethylene.

| Gas or vapour | Units | Permeability | |
| | | LDPE | HDPE |
| --- | --- | --- | --- |
| Oxygen | $cm^3/m^2/24$ h/atm (for 25 µm film) | 6500–8500 | 1600–2000 |
| Carbon dioxide | $cm^3/m^2/24$ h/atm (for 25 µm film) | 30 000–40 000 | 8000–10 000 |
| Water vapour | $g/m^2/24$ h (for 25 µm film at 90% RH and 38 °C | 15–20 | 5 |

an antioxidant. Reports have been published of LDPE, protected in this way, being serviceable after six or more years in the tropics.

One phenomenon that must be taken into account when considering the chemical resistance of polyethylene under conditions of stress is environmental stress cracking. When polyethylene is stressed multiaxially, in contact with certain polar liquids or vapours, cracking may occur. The substances that cause this stress cracking may not be solvents for, or even be more than slightly absorbed by, the polyethylene and if the polyethylene sample is unstressed no cracking occurs in contact with these very same substances. The probability of stress cracking occurring decreases with decreasing melt flow index (increasing molecular weight). An account of a test for environmental stress crack resistance is given in Chapter 22.

*Table 2.5* Classification of some materials as stress cracking agents for polyethylene.

| Active | Inactive |
|---|---|
| Metallic soaps | Water |
| Sulphonated and sulphated alcohols | Sugars |
| Polyglycol esters | Inorganic salts (acid and neutral) |
| Liquid hydrocarbons | |
| Silicone fluids | Polyhydric alcohols |
| Organic esters | Rosin |

Table 2.5 gives a classification of certain materials as stress cracking agents for polyethylene.

### 2.1.3 Linear low-density polyethylene

Linear low-density polyethylene (LLDPE) is a more recent development than either LDPE or HDPE. It has a molecular structure similar to that of HDPE (but with slightly longer and more numerous side chains) and a density similar to that of LDPE. As with the other polyethylenes, there are gas phase, liquid/slurry and liquid/solution processes. The properties of LLDPE, will vary according to the process employed, in the same way that the properties of LDPE and HDPE vary.

The major difference between LLDPE and LDPE is that the former's molecular weight distribution is narrower. In general, the advantages of LLDPE over LDPE are improved chemical resistance, improved high- and low-temperature performance, greater stiffness, higher surface gloss and a higher strength at a given density. LLDPE also has a greater resistance to environmental stress cracking. Some typical values are given in Table 2.6.

The differences in molecular structure affect the rheology of the two materials. LLDPE is more viscous at extrusion shear rates and requires greater power to extrude. In addition, a wider die gap is required when extruding 100% LLDPE in order to prevent melt fracture. For best results, some minor modifications have to be made to the screw and die gap of extruders designed for ordinary LDPE. Equipment has been designed to handle LLDPE, while another approach is the use of additives that allow the extrusion of LLDPE without the occurrence of melt fracture. Work

*Table 2.6* Environmental stress crack resistance of LDPE and LLDPE.

| Material | Melt flow index | Density (g/cm$^3$) | Hours to 50% failure |
|---|---|---|---|
| LDPE | 4 | 0.92 | 30 |
| LLDPE | 4 | 0.92 | 600 |

is also proceeding on the development of LLDPE grades that can be extruded through conventional LDPE equipment without modification.

The combination of improved properties has led to LLDPE making inroads into many film markets because similar performances can be obtained at lower thicknesses (downgauging). In injection moulding, improved stiffness, improved heat distortion resistance and higher environmental stress crack resistance can be coupled with higher melt flow indices with resulting faster cycle times. In blow moulding, too, LLDPE offers greater stiffness and environmental stress crack resistance compared with LDPE.

More recent developments include the production of so-called very low and ultra low density polyethylenes (i.e. below 0.915 g/cm$^3$). Examples of very low density polyethylenes (VLDPE) include Attane from Dow Chemicals (0.912 g/cm$^3$), Norsoflex from Orkem and another from Sumitomo Chemicals. Union Carbide offer a range of ultra low density polyethylenes (ULDPE) with densities from 0.900–0.905 g/cm$^3$.

Physical properties for these polymers are said to be superior to those of linear low density polyethylene with higher elongation, better puncture resistance, hot tack and optical properties and an enhanced water vapour barrier performance.

## 2.2 Polypropylene

Early attempts to polymerise propylene using high-pressure gas-phase processes gave only oily liquids or rubbery solids of no commercial value. Later work, using liquid-phase processes and catalysts similar to those used in the Ziegler HDPE process, gave crystalline, high molecular weight polymers of propylene. The particular catalyst used controls the position of each monomer unit as it is added to the growing polymer chain, thus allowing the formation of a polymer of regular structure from the asymmetric propylene monomer unit. This regularly structured polypropylene is known as 'isotactic' while the haphazardly arranged material is known as 'atactic'. The regular structure of isotactic polypropylene allows close approach of the polymer molecules, and crystallisation can then occur in a similar manner to that of polyethylene.

### 2.2.1 Physical properties

Polypropylene is colourless, odourless, has a low density (0.90 g/cm$^3$) and a relatively high softening point (150 °C). The physical properties of polypropylene depend mainly on:

(a) percentage of atactic material present

(b) molecular weight
(c) molecular weight distribution
(d) crystallinity

Most commercially available polypropylenes have approximately the same isotacticity while the crystallinity of a sample of given isotacticity depends on its thermal history. The main differences between commercial grades of homopolymers lie, therefore, in their molecular weight and molecular weight distribution. As with polyethylene, the molecular weight is commonly denoted by the melt flow index. In the case of polypropylene the determination is carried out at 230 °C instead of 190 °C in order to allow for the higher melting point of polypropylene.

High molecular weight polymers (low MFI) are obviously very viscous when melted and are difficult to injection mould or extrude through restrictive dies. They do, however, possess good toughness and good melt strength. Materials with a narrow molecular weight distribution do not generally have high melt strengths but they are more easily processed so that higher molecular weight grades (with greater toughness) can be used without sacrificing processability.

The general physical properties of polypropylene are similar to those of HDPE. Properties that are significantly better in polypropylene are hardness (HDPE, 65–70; polypropylene, 75—Shore D) and softening point (HDPE, 122–130 °C; polypropylene, 140–150 °C), while it has a slightly lower mould shrinkage and gives mouldings with a higher gloss.

In thin sections, polypropylene has a very high resistance to repeated flexing. It is thus possible to mould articles with built-in hinges. Integrally moulded hinges have been produced that have withstood over one million flexes without failure; even at temperatures down to −70 °C, correctly designed and moulded hinges survive over 50 000 flexes, while at −30 °C the figure would be over 200 000.

The impact strength of polypropylene is lower than that of HDPE although still adequate for most purposes. The impact strength is affected by temperature and there is some reduction in that of polypropylene at temperatures below about 0 °C. The impact strength at low temperature can be improved by the addition of small quantities of butyl rubber or poly(isobutylene) although this entails some sacrifice in rigidity and in resistance to high temperatures. Another method of achieving higher impact strength at low temperatures is by copolymerisation of propylene with other olefins. The subject of copolymers will be dealt with in Section 2.2.3. Polypropylene also has a high resilience and this is made use of in the design of wadless closures (see Appendix B).

The presence of the bulky methyl groups along the polypropylene chain leads to an increased resistance to creep over that displayed by LDPE or

HDPE. This is particularly noticeable at temperatures above 100 °C and under high dead-weight conditions such as those encountered at the bottom of large stacks of filled bottle crates.

Polypropylene is subject to oxidative degradation at the elevated temperatures necessary for processing but the addition of small amounts of an antioxidant is able to stabilise it during processing and at elevated service temperatures. For photo-oxidative degradation, the incorporation of about 2% of finely divided carbon black is satisfactory.

One of the uses for polypropylene is the injection moulding of thin-wall pots and it is interesting to note the type of grade development that can be achieved. In 1970, the minimum wall thickness for a 5 oz. yoghurt type pot was 0.508 mm (0.020 in.) but in 1980 this had been reduced to 0.356 mm (0.014 in.). Moulding times were reduced from 6.5 s to 3.8 s while the weight of the pot was reduced from 8 g to 5 g.

The use of fillers in polypropylene can greatly increase its stiffness, improve its processing behaviour and change its appearance. Commonly used fillers include talc, glass fibre, mica, cellulose fibres and carbon black. In general, toughness is reduced, while density and cost/unit volume are increased.

### 2.2.2 Chemical properties

The chemical resistance of polypropylene is similar to that of polyethylene although here again it possesses the additional advantage of being usable at higher operating temperatures. It has better resistance to oils and greases than has polyethylene and is not subject to environmental stress cracking. The permeability to oxygen, carbon dioxide and water vapour lies somewhere between those of LDPE and HDPE but it should be noted that biaxial orientation of polypropylene reduces its permeability.

### 2.2.3 Copolymers

The most commonly used comonomer is ethylene, usually at levels of between 1% and 5% by weight. Most useful materials result when the ethylene is distributed randomly along the polypropylene chain. These 'random copolymers' have good clarity and have lower and broader melting points than the homopolymers. Larger amounts of ethylene are used as comonomer to give 'impact copolymers'. These are much tougher at low temperatures, are less stiff and are relatively opaque.

The process of stretch blow moulding (see Chapter 11) has opened up new markets for polypropylene, both for homopolymers and for random copolymers. Table 2.7 shows some of the physical properties of a typical homopolymer and random impact copolymers.

*Table 2.7*  Typical properties of polypropylene.

| Property | Units | Homopolymer | Random copolymer | Impact copolymer |
|---|---|---|---|---|
| Melt flow index | g/10 min | 4 | 6 | 4 |
| Tensile strength at yield | M N/m$^2$ | 34.4 | 27.6 | 26.9 |
|  | (psi) | (5000) | (4000) | (3900) |
| Elongation at yield | % | 11 | 14 | 13 |
| Impact strength (Izod) | J/m | 43–54 | 107 | 133 |
|  | (ft lb/in.) | (0.8–1.0) | (2.0) | (2.5) |

## 2.3 Ethylene/vinyl acetate copolymers (EVAs)

EVA copolymers containing 7–8% of vinyl acetate can be considered as modified LDPE, while those with around 15–28% behave more like flexible PVC.

In general, EVA differs from LDPE in the following properties having:

(a)  higher impact strength
(b)  greater elasticity
(c)  higher permeability to water vapour and gases
(d)  better low-temperature properties
(e)  higher filler retention
(f)  greater flexural life
(d)  greater resistance to environmental stress cracking

Although it is mainly used for the manufacture of films, applications have included snap-on caps and liners for bag-in-box containers.

## 2.4 Ionomers

The term 'ionomers' refers to a family of polymers in which there are ionic forces between the polymer chains, as well as the usual covalent bonds between the atoms making up each chain. Basically they are ethylene/methacrylic acid copolymers with some of the acid groups present as a metal salt. The metal ions provide the cationic part of the link while carboxyl groups, located along the polymer chain, constitute the anionic portion.

The ionic interchain links strengthen and stiffen the polymer but unlike the covalent cross-links present in thermosets, they do not destroy its melt processability. Ionomers do have greater melt strength than that of polyethylene itself, and they are, therefore, very suitable for extrusion coating, skin packaging and vacuum forming. The ionic interchain forces also have an appreciable effect on crystalline morphology. Whereas LDPE itself is a semi-crystalline polymer and can only be given good clarity by rapid quenching (with a consequent inhibition of spherulite growth), the ionic forces in ionomers eliminate all traces of visible spherulites, so giving excellent clarity.

Ionomers have a greater resistance to oils and greases than has LDPE at room temperature, although the differences are not so marked at elevated temperatures. Chemically they are resistant to weak and strong alkalis but they are slowly attacked by acids. Environmental stress crack resistance is good and is claimed to be higher than that of LDPEs of equivalent melt flow index. Other chemical properties of ionomers are similar to those of LDPE, being resistant to alcohols, ketones and esters but swelling slightly in hydrocarbon solvents. The use of carbon black as an antioxidant is also necessary for outdoor applications.

Gas permeability of ionomers is similar to that of LDPE but water vapour permeability is somewhat higher, as is water absorption. The polar nature of ionomers means that printing is easier than on LDPE but some pretreatment is still necessary.

Transparent vials with integrally hinged caps have been moulded from ionomer but, again, the majority of uses lie in the fields of films and coatings.

# CHAPTER 3

# Styrene polymers and copolymers

Polystyrene is a well-established polymer and is one of the major thermoplastics. It is manufactured by the polymerisation of styrene in the presence of heat and is completely amorphous. This is because the bulky nature of the benzene rings in the polymer chains does not permit their close approach. In its unmodified state, polystyrene is a crystal-clear, hard, rigid polymer. It has many excellent properties but for some applications its impact strength is not always adequate. High-impact grades are produced by incorporating synthetic rubbers, usually by polymerisation of a rubber-in-styrene solution.

## 3.1 Unmodified or crystal polystyrene

Crystal polystyrene is a colourless, transparent thermoplastic with a distinctly metallic ring when dropped on to a hard surface. It is hard, has a fairly high tensile strength and a high refractive index. It softens at about 90–95 °C, while at 140 °C it is a mobile liquid, excellent for injection moulding. Intrinsically it is a rather brittle material and, as mentioned earlier, toughened (or high-impact) grades have been developed for the more critical applications.

A large number of grades of polystyrene have been manufactured for specific purposes. In addition to the high-impact grades already mentioned, there are such crystal grades as general purpose, quick processing and heat resistant. These grades are obtained by a combination of varying molecular weights and molecular weight distributions, coupled with the addition of varying amounts of lubricant. Among the lubricants in general use are such compounds as esters, organic acids and mineral oils. In addition to improving the flow of the polystyrene, these lubricants also reduce the brittleness, lower the softening point and increase the setting-up time. The presence of low molecular weight polystyrenes also has the effect of improving the flow but in this instance the brittleness is increased.

Quick processing grades can be produced either by improving the flow properties or by reducing the setting time. Resins with reduced times usually contain very little lubricant and may be of slightly lower molecular weight than that of a general purpose grade. This lower molecular weight confers slightly improved flow characteristics as an additional aid to speedy processing. On the other hand, the improved flow grades contain fairly large amounts of lubricant and a large proportion of low molecular weight polystyrenes. These low molecular weight polymers increase the brittleness but this is counterbalanced by the amount of internal lubricant used.

Heat-resistant grades are of high molecular weight, contain no internal lubricants and have a minimum amount of residual styrene monomer or low molecular weight products. This reduces the flow properties and makes injection moulding more difficult. Moulded articles are often annealed in order to relieve any stresses that may have occurred during moulding. This is carried out quite simply by placing the mouldings in warm water immediately on removal from the mould.

Chemically, polystyrene is resistant to strong acids and bases and is insoluble in aliphatic hydrocarbons and the lower alcohols. It is, however, soluble in esters, aromatic hydrocarbons, higher alcohols and chlorinated hydrocarbons.

Polystyrene is a poor barrier to water vapour but its gas barrier properties are slightly better than those of LDPE. It is a good barrier to essential oils and perfumes which makes it a good choice for the packaging of cosmetics and scented products generally.

## 3.2 High-impact polystyrene

The increase in impact strength of high-impact polystyrenes is achieved at the expense of a loss in clarity and such grades are either translucent or opaque. High-impact grades have higher elongation at break, reduced softening points, reduced tensile strengths and reduced surface hardness. The increased flexibility of high-impact polystyrenes allows the use of metal inserts in mouldings. This is because the flexibility compensates for the differences in the coefficients of expansion between the plastic and the metal.

As with polypropylene (see Chapter 2), grade development is an ongoing process. In 1970, a typical medium-impact polystyrene gave a minimum wall thickness for a 5 oz. yoghurt pot of 0.635 mm (0.025 in.); this was reduced by 1980 to 0.406 mm (0.016 in.), moulding times were reduced from 5 s to 3 s while pot weights were reduced from 12 g to 6.5 g.

Chemically, high-impact polystyrenes have similar properties to those of crystal polystyrene. The properties of some typical crystal and high-impact polstyrenes are given in Table 3.1.

Table 3.1  Typical properties of polystyrene.

| Property | Units | Unmodified polystyrenes | | | Toughened polystyrenes | |
|---|---|---|---|---|---|---|
| | | General purpose | Quick processing | Heat resistant | High impact | Extra high impact |
| Density | g/cm$^3$ | 1.05–1.07 | 1.05–1.07 | 1.05–1.07 | 1.05–1.07 | 1.04–1.06 |
| Tensile strength | psi | 6000–7500 | 6000–7500 | 7000–8000 | 4500–5500 | 2500–3500 |
| | (M N/m$^2$) | (41–52) | (41–52) | (48–55) | (31–38) | (17–24) |
| Elongation | % | 1–2.5 | 1–2.5 | 1.5–2.0 | 12.5–22.5 | 25–45 |
| Impact strength | ft lb/in. | 0.2–0.35 | 0.2–0.35 | 0.2–0.35 | 0.6–1.0 | 1.0–2.0 |
| (Izod) | (J/m) | (11–19) | (11–19) | (11–19) | (33–54) | (54–108) |
| Softening point | °C | 85–90 | 82–84 | 100–103 | 85–90 | 78–83 |

### 3.3  Expanded polystyrene

Polystyrene is also available in the form of expandable beads that can be processed by steam moulding to give light-weight products with many uses in the fields of packaging and thermal insulation. The expandable beads can be produced either by polymerising and gassing in one step, or by polymerising the styrene, then impregnating the resultant beads at elevated temperatures and pressures. In either case the gassing agent is pentane.

The moulding of the expandable polystyrene beads is described in Chapter 14.

### 3.4  Acrylonitrile/butadiene/styrene (ABS)

ABS is a graft copolymer of styrene and acrylonitrile on to a polybutadiene rubber backbone. Like high-impact polystyrenes, ABS is translucent to opaque. ABS polymers have high-impact strength and are extremely tough and hard.

Chemically, ABS is resistant to alkalis and dilute acids. Concentrated acids cause slight swelling and darkening. It is also resistant to alcohols and to detergents, animal fats and vegetable oils. Esters, however, severely attack ABS and methyl ethyl ketone causes disintegration.

It is easily thermoformed and has been used in the manufacture of margarine tubs. This sort of application could well have been suitable for high-impact polystyrene but ABS was chosen because of its superior resistance to stress cracking. It is also used for the injection moulding of transit trays because it warps less than, say, HDPE and polypropylene.

# CHAPTER 4

# Vinyls

For the purpose of this book the vinyl family of polymers is taken to include poly(vinyl chloride), vinylidene chloride copolymers and vinyl alcohol copolymers.

## 4.1 Poly(vinyl chloride) (PVC)

PVC is a hard, brittle material having a density of around 1.35–1.4 g/cm$^3$. This basic material can be modified, by the addition of plasticisers, to give soft and flexible products. Plasticisers are chemicals that are soluble in PVC and act as internal lubricants by increasing the distance between the polymer chains with a consequent decrease in interchain attractive forces. In addition to being soluble in PVC, plasticisers should possess low volatility (to improve retention), be colourless and (when used in contact with foodstuffs) be free from toxic hazard or taint.

PVC normally starts to degrade at temperatures not much higher than normal processing temperatures. The problem is eased by streamlining processing equipment internally to facilitate flow and avoid hold-up of material in the heated cylinder. Plasticisation also facilitates flow through the processing equipment and allows processing temperatures to be reduced. Where rigid formulations are required, heat stabilisers can be added. Special grades of plasticisers and heat stabilisers are available for food-contact uses.

Unplasticised PVC has excellent resistance to oils, greases and fats and is also resistant to acids and alkalis. It is softened by certain solvents however, particularly ketones and chlorinated hydrocarbons. The water vapour barrier properties of PVC are poorer than those of polypropylene and the polyethylenes but are still adequate for many purposes. Gas barrier properties are higher than those of the polyolefins, however, and PVC bottles can be used for the packaging of oils and fats.

It is difficult to be specific about the physical properties of PVC because of the wide range of plasticisation possible. In general, increases in plasticiser content lower the tensile strength, increase the elongation at breaking point and improve the low-temperature properties. Embrittlement can, however, occur in outdoor uses due to loss of plasticiser. Most PVC bottles are made from unplasticised or lightly plasticised grades and impact strength has to be improved by the use of impact modifiers.

Impact modifiers are probably the most important additives in a PVC blow moulding compound. They include ABS copolymers, methyl methacrylate/butadiene/styrene copolymers (MBS), EVA and acrylic resin derivatives. The final choice depends not only on the impact resistance required but also on cost, and on any possible deterioration in transparency or odour. Additions of 5–20% are usual, depending on the type of modifier used and on the impact-strength improvement required. Correct design of the bottle also has a considerable effect on the impact resistance. Sharp corners should be avoided and wall thickness distribution should be as uniform as possible.

PVC also benefits from orientation, and stretch blow moulded PVC bottles have a high resistance to breakage at minimum unit weight, a high pressure resistance and excellent optical properties (brilliance, absence of streakiness and high transparency). The greater strength imparted by orientation allows lower-weight bottles to be made with satisfactory impact strength. One example quoted is a 0.33 litre bottle to hold carbonated drinks (at 6 vols carbonation) where a 25 g biaxially oriented bottle is said to be equivalent to a 40 g normal extrusion blow moulded one. Orientation also allows a great reduction to be made in impact modifier content. In fact, 30% higher drop strengths have been claimed with 20% reduction in weight and with no impact modifiers added. In addition to the cost factor, this is important in the packaging of disinfectants. This is because they normally attack and cause discolouration of impact-modified grades of PVC.

## 4.2 Vinylidene chloride copolymers

Vinylidene chloride is copolymerised with various other monomers, including vinyl chloride and methyl acrylate. Although the homopolymer is not used commercially, the name 'poly(vinylidene chloride)' (PVDC) is often used to cover the various copolymers. Copolymers are used because the homopolymer degrades rapidly at temperatures only about 5–10 °C above the melting point. Copolymerisation brings the melting point low enough to render processing feasible.

Although PVDC is not used as a blow moulding material on its own, it is of interest (a) as a coating material for bottles, to improve their barrier properties and (b) as a component in coextruded bottles (for the same reason).

### 4.2.1 Properties

Physical and chemical properties depend largely on the vinylidene chloride content which ranges from about 72% to 92% by weight. The higher the vinylidene chloride content, for example, the higher the barrier properties to water vapour and gases. The degree of crystallinity also relates directly to barrier properties. A typical PVDC film would have an oxygen permeability of 10 cm$^3$/m$^2$ day (for 25 μm thickness) which compares with LDPE (7000) and HDPE (2000).

## 4.3 Ethylene/vinyl alcohol copolymers

Ethylene/vinyl alcohol copolymers (EVOH), like PVDC, are of interest as components of coextruded bottles. They are highly crystalline, with properties that are very much dependent on the relative concentrations of the comonomers. An idea of the range of properties available is given in Table 4.1.

The most outstanding characteristic of EVOH copolymers is their ability to provide a barrier to gases such as oxygen, carbon dioxide and nitrogen. The oxygen barrier properties, in particular, are important in preserving flavour and quality retention.

EVOH copolymers are hydrophilic and their high gas and solvent barrier properties are reduced under humid conditions. However, because the barrier properties are so high, it is feasible to use the material in very low gauges and it is coextruded, with other polymers, such as HDPE and polypropylene, on the outside of the multilayer structure, so giving protection to the EVOH and reducing the cost.

EVOH copolymers have high mechanical strength, elasticity and surface hardness. They are also highly resistant to abrasion. Resistance to oils and solvents is very high. The percentage increase in weight of EVOH has been measured after one year's immersion in various solvents and oils. Figures were zero for benzene, xylene, petroleum ether, acetone and cyclohexane, 0.12% for salad oil, 2.3% for ethanol and 12.2% for methyl alcohol. This makes EVOH an excellent choice for the packaging

*Table 4.1*  Range of EVOH resins.

| Property | Range |
|---|---|
| Melt index | 0.7–20 |
| Density (g/cm$^3$) | 1.13–1.21 |
| Ethylene content (mol %) | 29–48 |
| Melting point (°C) | 158–189 |
| Oxygen permeability (cm$^3$/m$^2$/24 h/atm) (for 25 μm film) | 0.16–1.46 |
| | (PVDC = 2.40) |

of oily foods, edible and mineral oils, and organic solvents. HDPE/EVOH combination containers have been produced in sizes up to 5 litres and are said to be suitable for petrol, liquid fertilisers and herbicides.

Semi-rigid containers are also produced from coextruded sheet containing EVOH as the barrier layer. Among the structures used are high-impact polystyrene/tie layer/EVOH/tie layer/polypropylene, nylon/EVOH/nylon/ionomer and polypropylene/tie layer/EVOH/tie layer/polypropylene.

# CHAPTER 5

# Polyamides (nylons)

## 5.1 Characterisation

Nylons are made either by the condensation of a diacid and a diamine or by the condensation of certain $\omega$-amino acids. In the first instance the resultant polymer is characterised by a number derived from the number of carbon atoms in the parent diamine and diacid. In the latter instance the number is derived from the number of carbon atoms in the parent amino acid. Thus, nylon 6,6 is made by condensing hexamethylene diamine and adipic acid (both substances having six carbon atoms in the molecule) and nylon 11 is produced from $\omega$-undecanoic acid, which has eleven carbon atoms in the molecule.

## 5.2 Physical properties

Nylons, generally, are produced in the form of hard, horny chips, creamy in colour and slightly translucent. The various types of nylons can vary in their properties but the following generalisations can be made for nylons 6, 11, 6,6 and 6,10.

Nylons are tough, have a high impact strength, high tensile strength and a marked resistance to abrasion. They are highly crystalline and because of this they possess sharp melting points that are appreciably higher than many other thermoplastics. They also exhibit very little cold flow. Nylons are slightly hygroscopic and the material must be dried before processing. The mechanical properties of nylons are reduced somewhat by the absorption of water but the effect is not permanent and the full mechanical properties are restored when the material is dried. Because of their polar nature, nylons can be printed without pretreatment.

*Table 5.1*   Effect of substitution in the parent diacid or diamine on the melting point of nylon 6,6.

| Dibasic acid | Diamine | Melting point (°C) |
|---|---|---|
| Adipic acid | Hexamethylene diamine | 265 |
| Methyl adipic acid | Hexamethylene diamine | 166 |
| Adipic acid | 3-Methyl hexamethylene diamine | 180 |

*Table 5.2*   Properties of a range of polyamides.

| Property | Units | Type | | |
|---|---|---|---|---|
| | | 6,6 | 6 | 11 |
| Density | g/cm$^3$ | 1.14 | 1.13 | 1.05 |
| Tensile strength | psi | 11 000 | 12 000 | 8 500 |
| | (M N/m$^2$) | (76) | (83) | (59) |
| Elongation | % | 90 | 300 | 120 |
| Water absorption | % | 1.5 | 1.6 | 0.4 |
| Melting point | °C | 265 | 215 | 185 |
| Oxygen permeability | cm$^3$/m$^2$/24 h/atm | 55 | 45 | 200 |
| | (for 25 μm film) | | | |
| Water vapour permeability | g/m$^2$/24 h | 280 | 270 | 80 |
| | (for 25 μm film at 90%RH and 38 °C | | | |

Copolymerisation tends to inhibit crystallisation because it breaks up the regular structure along the polymer chain. The melting points are, consequently, lower. Copolymers are also generally more soluble, less stiff and softer than the homopolymers. The effect on polymer melting point brought about by using substituted diamines or dibasic acids is shown in Table 5.1.

Some typical values for the physical properties of various nylons are given in Table 5.2.

## 5.3  Chemical properties

Chemically, nylons are inert to inorganic reagents although they are attacked by oxidising agents such as hydrogen peroxide and chlorine bleaches. Aqueous solutions of inert inorganic chemicals still exert some effect on the physical properties of moulded nylon specimens, however, because of the plasticising effect of the water itself. Nylons are attacked by concentrated mineral acids at room temperature but dilute acids have little effect except when hot.

Organic acids have little or no effect (especially in low concentrations), notable exceptions being 90% formic acid and glacial acetic acid which cause swelling and solution. Nylons are resistant to caustic alkalis, even at concentrations up to 20%, while they are even more stable towards other alkaline solutions such as trisodium phosphate, sodium carbonate and potassium carbonate.

Nylons are of particular interest because of their resistance to organic solvents. In general, they are resistant to alcohols, ether, acetone, petroleum ether, carbon tetrachloride, benzene, xylene and aviation oil. They are, however, attacked by phenols, hot formamide, formaldehyde and hot nitrobenzene. It should be realised, however, that certain copolymers or plasticised grades may be less resistant to certain chemicals. A terpolymer of 6, 6,6 and 6,10 nylons, for instance, is attacked by alcohols. Nylons are relatively unaffected by prolonged service indoors at room temperature but at elevated temperatures oxidation occurs with consequent yellowing and embrittlement, while strong sunlight also has an adverse effect on their mechanical properties.

Blow moulded bottles have been used to contain a wide range of chemicals and pharmaceuticals because of their chemical inertness. Other useful properties here, are flexibility, strength and sterilisability. Nylons are relatively expensive but their properties can be utilised economically by using them in coextrusion blow moulding. Nylon/HDPE and nylon/polypropylene coextruded bottles have given excellent results with respect to vitamin C retention when used to store orange juice. Other nylon/polyolefin coextruded bottles where nylon constitutes the inner layer are said to be useful for containing benzene, toluene, phenol and xylene. With the polyolefin layer inside, the bottles are said to be suitable for foodstuffs in aqueous solution form or for wines, liquors, etc.

## 5.4 Miscellaneous nylons

Normal nylons such as nylons 11, 6 or 6,6 are semi-crystalline and are sensitive to moisture. One example of this is the reduction in gas barrier properties under humid conditions. Amorphous nylons have now been developed (such as Du Pont's Selar PA) that have significantly higher moisture vapour barrier properties and have gas barrier properties that are *higher* under humid conditions.

Another development (by Mitsubishi Chemicals) is MXD6. This is actually a semi-crystalline poly(acrylamide) and has high barrier properties. One use in Japan is as the barrier core layer in a three-layer bottle for wine and beer. The outer and inner layers are both PET.

# CHAPTER 6

# Acrylics

## 6.1 Polyacrylonitrile

The homopolymer, polyacrylonitrile (PAN), has many properties that render it of interest as a bottle-blowing material. It is hard, has good clarity, is an outstanding barrier to gases and has extremely good resistance to a wide range of chemicals. Its impact strength and creep resistance are borderline but the impact strength can be improved by stretching and heat setting.

Unfortunately, it has one major handicap, namely, that it is not easily melt processable. To overcome this drawback, copolymers have been produced using various comonomers that improve melt processability while retaining the desirable properties already mentioned. One copolymer is ABS (described in Chapter 3) but the acrylonitrile content is too low for the gas barrier properties to be retained. Pure PAN contains 49% of the nitrile group (—CN) but high gas barrier copolymers with acceptable melt processability can be produced using nitrile contents over 25%. Such materials are usually referred to as high nitrile polymers (HNPs) or high nitrile resins (HNRs).

## 6.2 High nitrile polymers

### 6.2.1 Acrylonitrile/styrene copolymers (ANS)

An early example of an HNP was a copolymer combining acrylonitrile and styrene in a 70:30 ratio. Two companies, Monsanto and Borg-Warner, produced such polymers as materials for carbonated beverage bottles in the late 1970s. Later, in the USA, nitrile copolymers came under attack because of certain toxicological problems with acrylonitrile. The FDA then banned the use of high nitrile polymers in beverage packaging because of concern over the potential extraction of acrylonitrile from the bottle into the beverage. Such polymers were still allowed for some other uses, under certain conditions, but because the major commercial significance was for beverage packaging, commercial production was discontinued.

In 1984, as a result of petitions by the Monsanto Company, the FDA accepted that the extraction of acrylonitrile from an ANS bottle, the preform of which was irradiated prior to blowing, would be below the detection limit of 0.16 ppb. The FDA then ruled that acrylonitrile concentrations in the beverage at or below this detection limit would be considered acceptable. The limit for residual acrylonitrile in the finished container was set at 0.1 ppm.

### 6.2.2 Rubber-modified acrylonitrile/methacrylate copolymers

One example of a rubber-modified acrylonitrile/methacrylate copolymer is Barex, the trade name for a material developed by Sohio Chemical Company. It is made by copolymerising a 75:25 mixture of acrylonitrile and methyl acrylate in the presence of a small amount of a butadiene/acrylonitrile synthetic rubber. It has good clarity, excellent gas barrier properties and a high resistance to creep. In addition, it has good impact strength and is insoluble in a wide range of organic solvents.

Barex constituted part of the Rigello pack which was developed by Rigello Pak Ltd in Sweden for carbonated drinks. It was a novel container and was formed from four component parts. The bottom was a cylinder with a hemispherical base, moulded from Barex, which acted as a liquid and gas barrier. Mechanical strength was supplied by a sleeve wound from three layers of paper. The design was carried on the outer layer, which was then coated with LDPE, or an outer liner of aluminium foil. Barex also constituted the conical top which was welded to the base component. The cone had an orifice into which was fitted a cap, injection moulded from medium-density polyethylene. The inside of the cap was coated with PVC to reduce gas transmission.

Rigello was used for a number of years in Sweden for the packaging of beer and, for a short period, in the UK. It was phased out for economic reasons. Although it was, at one time, competitive with its main rival, the can, prices of the latter went down and when there was a devaluation of the Swedish Krona the acrylic copolymer became too expensive to compete. Nevertheless, the Rigello pack remains as an example of how the correct use of a number of materials can produce a solution to a problem that was insoluble by any one plastics material (at that time).

Rubber-modified HNPs are used in thermoforming and in bottle blowing. Injection blow moulding is the preferred method for small containers such as those used for typing correction fluid, nail polishes and other cosmetics. Larger bottles are normally made by extrusion blow moulding. Applications here, include agricultural chemicals, fuel additives and aggressive household chemicals. Extrusion/stretch blow moulding is also gaining ground for the manufacture of medium to large bottles. The orientation obtained in this process increases the impact strength and allows thinner walled containers to be used.

Rubber-modified HNPs are also used in coextrusions, usually with polyolefins such as HDPE or polypropylene. A typical structure is HDPE/tie layer/RM-HNP with the barrier polymer in direct contact with the aggressive contents of the bottle.

# CHAPTER 7

# Poly(ethylene terephthalate) and copolymers

Poly(ethylene terephthalate), better known as PET, was patented as a fibre-forming polymer by J. R. Whinfield and J. T. Dickson while working in the laboratories of the Calico Printers Association in 1941. The patent rights were later acquired by Imperial Chemical Industries (ICI) and E. I. Du Pont de Nemours, who eventually produced the world-famous polyester fibres, 'Terylene' and 'Dacron', respectively. Later still, film grades were developed so that a large manufacturing base already existed for PET when stretch blow moulding was introduced. PET homopolymers could not at that time be processed by normal extrusion blow moulding techniques because of insufficient melt strength. New grades and improved machine controls have since made this practicable.

PET was first used in the stretch blow moulding of carbonated beverage bottles during the 1970s. This market grew from 70 000 tonnes to 300 000 tonnes in 1985 in the USA alone. Similar growth occurred in many other countries (though from a later start) and the estimated tonnage world-wide for 1985 was 700 000 tonnes. Estimates for future growth include a figure of 615 000 tonnes for Western Europe alone by 1995.

## 7.1 Manufacture

The manufacturing process for PET is of interest because of its influence on the final end-use properties. Two different routes are available, namely, from terephthalic acid (TPA), by esterification, and from dimethyl terephthalate (DMT), by ester interchange, both materials being reacted with ethylene glycol (EG). The route using DMT was favoured because it was more easily purified than was TPA. However, the problems of purifying TPA have now been overcome and the TPA route is now favoured because it is the simpler one chemically. The chemistry of PET manufacture from terephthalic acid is briefly as follows.

The initial reaction is esterification of the acid with ethylene glycol to give the monomer diethylene glycol terephthalate (DGT) plus water. The DGT is then polycondensed to yield a long chain polymer containing around 100 repeat units. Whereas the initial esterification reaction proceeds with the elimination of water as a by-product, the polycondensation step (which proceeds under the influence of heat and under a high vacuum) releases ethylene glycol. This method is the one used by ICI plc in the manufacture of their bottle-blowing grade Melinar.

$$COOH \quad + \quad 2(CH_2OH)_2 \quad \rightarrow \quad COO.CH_2.CH_2OH \quad + 2H_2O$$

|  Terephthalic acid  |  Ethylene glycol  |  Diethylene glycol terephthalate  |
|---|---|---|

*heat and reduced pressure*

$$HO.CH_2-(CH_2.O.C-\!\!\!\bigcirc\!\!\!-CO.CH_2)_n-CH_2OH + (n-1)(CH_2OH)_2$$

Poly(ethylene terephthalate)

Ethylene glycol

For bottle-blowing purposes, the PET pellets or chips, as produced by the process just outlined, have certain drawbacks. These are:

1. the chips are amorphous
2. they are high in retained acetaldehyde content, and
3. they have a low molecular weight.

In addition to acetaldehyde there are other impurities that might promote degradation during subsequent processing. The amorphous polymer also tends to fuse and form lumps in the drying hopper. It is upgraded, therefore, by crystallising the chips to avoid sticking and then drying to reduce hydrolysis at high processing temperatures. The process is a solid-phase polymerisation and can be a batch one (high temperature under vacuum) or continuous (high temperature in a stream of nitrogen). The process improves drying and moulding characteristics, and bottle production and quality.

## 7.2 Physical properties

When oriented, as in stretch blow moulding, PET has a high tensile strength and is well able to resist the pressures generated by normal carbonated soft drinks ($345-415$ kN/m$^2$; $50-60$ lb/in.$^2$). Gas barrier properties are higher than those of PVC, polycarbonate and the polyolefins but are not as good as those of poly(acrylonitrile). It is also a good barrier to water vapour.

The softening point is high, the material melting over a range of $243-270$ °C. It has a high thermal and thermo-oxidative stability and has good sparkle and optical clarity (comparable to that of glass).

Although gas barrier properties are good, there are areas where improvements in carbon dioxide retention and resistance to oxygen ingress would open large areas of business, such as the packaging of beer. It is known that PET bottles of 1 litre or above can retain carbon dioxide to an acceptable extent (not more than 15% lost over about four months) but as bottle size is reduced, the increase in surface area to volume ratio makes this limit difficult to achieve. In addition, beer is adversely affected by traces of oxygen. Ingress of oxygen in small bottles of PET is sufficient to make them unacceptable for beer bottling. Better gas barrier properties can be achieved by three main methods, namely,

1. by modifying the PET polymer to give inherently better barrier properties,
2. by using thicker walled bottles, or
3. by producing multilayered bottles via dip coating, coextrusion or co-injection moulding.

The first solution is the neatest, and work has been carried out and is continuing in this area. The use of thicker walled bottles suffers from the disadvantage that carbon dioxide is lost not only by permeation but by solution in the bottle wall. There is an equilibrium between the two that limits the thickness that can be employed.

As far as multilayered bottles are concerned, there have been a number of approaches. One is the coating of a stretch blow moulded PET bottle with a poly(vinylidene chloride)-based coating. Such coated bottles have increased barrier properties with respect to both carbon dioxide loss and oxygen ingress and have been used commercially for the packaging of both beer and wine. Co-injection stretch-blowing machines have also been developed using, for example, PET/acrylonitrile copolymer and PET/ EVOH/PET bottles.

The viscosity of PET is important because it directly controls the minimum or natural draw ratio that has to be applied to obtain a nearly uniform wall thickness. Resins of relatively low viscosity (intrinsic viscosity

(IV) less than 0.75) may be used for the stretch-blow moulding of large bottles (2 litres or more) but for smaller bottles 1.5 litres or less) higher viscosities are preferred.

The orientation of PET induces crystallinity, fixing the polymer chains into ordered positions. When the bottles are heated to temperatures of around 78 °C (the glass transition temperature of PET) the polymer chains vibrate and the bottles tend to distort. PET bottles are normally unsuitable, therefore, for steam sterilisation or for hot filling above 75 °C. Annealing at temperatures above 140 °C overcomes this but there are problems such as loss of clarity and of some mechanical strength in the thicker areas of the bottle. Special bottle-blowing equipment has been developed using hot moulds to heat-set the bottles, and such bottles are said to be hot fillable at temperatures up to 85 °C. It is possible to produce bottles stable at 100 °C but higher cycle times are necessary to allow equilibrium to be reached and this increases the cost.

Other heat-stabilising techniques that allow the hot filling of PET are:

(a) crystallising the neck zone, and
(b) the use of in-mould polycarbonate neck inserts.

The catalyst system used in the polymerisation of PET has an effect on bottle clarity. Catalyst systems used are usually based on antimony, germanium or a mixture of antimony/germanium-containing compounds. All produce very clear bottles but germanium also adds an extra 'sparkle' to the bottle.

## 7.3 Chemical properties

PET is resistant to many chemicals, including hydrocarbons, chlorinated hydrocarbons, ketones, esters, concentrated hydrochloric acid and certain other dilute acids. It is, however, decomposed by alkalis and oxidising agents.

Although PET is an inert polymer, degradation of the molten polymer does occur and leads to the formation of acetaldehyde. In addition to its formation in such chemical reactions as the thermal degradation of PET, acetaldehyde is a natural product. It occurs in citrus fruits, grapes and grape products, apples and many other fruits and berries. It is also a natural constituent of butter, cheese, frozen vegetables and olives. Acetaldehyde has a characteristically pungent and penetrating odour and is used as a flavour and fragrance ingredient. It can be found in beverages, ice cream, chewing gum, rum, wine and chocolate. Its use as a food additive is officially approved and it is, therefore, a harmless chemical. However, it does possess a flavour and its concentration must be kept as low as possible, particularly in sensitive products such as colas and mineral waters. The specification set by cola manufacturers is 3 µg/litre. When PET degrades to form acetaldehyde and the PET preform is cooled, the acetaldehyde is entrapped in the solidified PET. It has been found that a concentration of less than 9 ppm acetaldehyde in the preform is necessary to achieve the specified 3 µg/litre in the cola.

Much work has been carried out to reduce the level of acetaldehyde in the finished PET bottle. Residence time of the molten polymer in the extruder or injection moulding machine is one factor, the generation of acetaldehyde being almost directly proportional to melt residence time. Another factor is temperature, but the picture here is not so simple. Processing temperatures are limited at the lower end of the range in terms of bottle clarity and at the upper end by the generation of acetaldehyde.

The polymer must be dried before moulding to a moisture content of 20 ppm to minimise hydrolytic degradation.

## 7.4 Crystalline PET (CPET)

Amorphous PET sheets are easily converted into a range of containers by thermoforming of the extruded sheet. Such containers have excellent clarity but are not suitable for 'ovenable' applications like trays for frozen food and prepared meals.

Ovenable trays are manufactured by thermoforming cast sheet with subsequent crystallisation. Crystallisable (CPET) grades contain nucleating agents which allow them to crystallise rapidly when thermoformed in a heated mould. Crystallisation heat sets the forming to prevent deformation during cooking and serving.

## 7.5 Copolymers

Copolymers can be made either by using more than one dibasic acid or by using more than one glycol. One dibasic acid used as a comonomer is isophthalic acid which is prepared from m-xylene. This gives resins with a higher than normal molecular weight. Such resins have superior strength characteristics, toughness and increased heat resistance. The use of isophthalic acid also reduces the rate and the degree of crystallisation. This, in turn, widens the processing range of the blow moulding equipment.

Glycols available as comonomers include neopentyl glycol, diethylene glycol and 1,4-cyclohexane dimethanol. The last named is manufactured and marketed under the trade name Kodar PETG 6763, by Eastman Chemical International A.G. Bottles blown from PETG have good clarity, toughness and chemical resistance. They are suitable for the packaging of shampoos, soaps, detergents and oils but are not recommended for carbonated beverages and other pressurised contents.

# CHAPTER 8

# Polycarbonate

Polycarbonate is a relatively expensive polymer and is manufactured from expensive starting materials. Polycarbonate does not have a sharp melting point but begins to soften above its glass transition point of 149 °C. It then goes into a rubbery state until between 200 and 220 °C, when it passes into the truly thermoplastic state.

## 8.1 Physical properties

Polycarbonate has a high impact strength, a high softening point and excellent clarity. It also retains these properties well with increasing temperature. At 125 °C, for example, polycarbonate film still exhibits a tensile yield strength of 31 MN/m$^2$ (4438 lb/in$^2$) which is about the same as that of HDPE at room temperature. Low-temperature properties are also excellent, the brittle point being below $-135$ °C. Some idea of polycarbonate's clarity can be obtained from its light transmission value which is 88–91%, as compared with 92% for clear plate glass. It also has a low haze value (less than 1%). Polycarbonate has a high gloss and a high resistance to staining to tea, coffee, inks, lipstick and many household materials.

## 8.2 Chemical properties

Chemically, polycarbonate is resistant to dilute acids but is strongly attacked by alkalis and bases, such as amines. It is resistant to aliphatic hydrocarbons, alcohols, detergents, oils and greases but is soluble in chlorinated hydrocarbons. Methylene chloride, for example, is used for the solvent cementing of polycarbonate. It is also partially soluble in aromatic hydrocarbons, ketones and esters. These materials will cause stress crazing at elevated temperatures or under conditions of stress. Water absorption is low and immersion in water at room temperature

Table 8.1 Typical properties of polycarbonate.

| Property | Units | Value |
|---|---|---|
| Density | g/cm$^3$ | 1.2 |
| Tensile strength | psi | 9000 |
| | (M N/m$^2$) | (62) |
| Elongation | % | 60–100 |
| Impact strength | ft lb/in | 12–16 |
| | (J/m) | (642–856) |
| Water absorption | % (24 h) | 0.3 |
| Water vapour transmission | g/m$^2$/24 h | 77–93 |
| | (for 25 µm film at | |
| | 90% RH and 38 °C) | |
| Oxygen permeability | cm$^3$/m$^2$/24 h/atm | 4500 |
| | (for 25 µm film) | |

for several weeks does not produce any appreciable change in tensile strength or elongation. However, immersion in boiling water does reduce elongation. In four hours the elongation drops from 100% to 50% although ultimate and yield tensile strengths are again unaffected.

Permeability to gases and water vapour is relatively high and if appreciable barrier properties are required then polycarbonate is co-extruded with, for example, EVOH. Some physical and chemical properties of polycarbonate are given in Table 8.1.

## 8.3 Uses

Before processing polycarbonate, the base polymer must be dried. Although polycarbonate possesses a relatively high melt viscosity, it can be processed on all conventional types of moulding machine. It can be cast or blow extruded into film and sheet, and injection moulded or blow moulded by both injection and extrusion blow moulding techniques. Polycarbonate sheet is readily thermoformable and gives quite deep draws with good mould detail. Polycarbonate's clarity and high strength have led to its use for returnable bottles, especially for 5 gallon water bottles in the USA. It has also been used in the USA for multi-trip, 1 gallon milk bottles.

Polycarbonate is a very useful material to incorporate into coextrusions. As well as supplying such properties as toughness, clarity and high-temperature resistance, polycarbonate acts as a support for other polymers with weaker melt strengths, so making for easier coextrusion.

Basically there are three processing areas where polycarbonate coextrusions can be used. The first is in thin multilayer film extrusion for use in flexible packaging. The second is thermoforming of thicker film or sheet for applications such as blister packaging, form/fill/seal containers for hot fill packaging, freezer-to-microwave oven-to-table packages and auto-clavable medical trays. Thirdly, there is the area of rigid packages, such as bottles and jars, for food packaging, cosmetics and pharmaceuticals.

*PART 2*

---

MANUFACTURING AND ANCILLARY PROCESSES

# Extrusion

Extruders are used to feed blow moulding machines that produce bottles. They are also used to produce sheet that can then be thermoformed to give trays, tubs, pots, etc. A common sequence of events is as follows:

1. plasticisation of the raw material (granules or powder),
2. metering of the plasticised product through a die that forms it to the desired shape,
3. solidification into the desired shape and size, and
4. winding into reels or cutting into units.

## 9.1 Construction

A typical extruder is shown in Fig. 9.1. It consists essentially of an Archimedean screw that revolves inside a close-fitting, heated cylinder or barrel. The thermoplastic material is fed into the extruder via a hopper. This is then carried along the barrel by the action of the screw and is

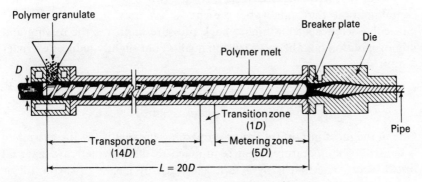

**Figure 9.1** Typical single-screw extruder. (Reproduced from Briston, *Plastics Films*, 3rd edition, Longman, London.)

gradually melted by contact with the heated barrel walls and by the generation of frictional heat by the screw. The final section of the screw forces the polymer melt through a die that determines the final form. An annular die extrudes a tube while a slit die extrudes flat film or sheet.

### 9.1.1 Screw design

The successful operation of any extruder depends on the design of the screw. It is often found to be impossible to extrude one material satisfactorily using a screw designed for some other material. Screws are characterised by their length/diameter ratios ($l/d$ ratios) and by their compression ratios. The compression ratio is the ratio of the volume of one flight of the screw at the hopper end to the volume of one flight at the die end.

The screw is normally divided into three sections – feed, compression and metering – but there may be further subdivisions. The feed section conveys the material from under the mouth of the hopper to the compression section, where the gradually diminishing depth of thread causes volume compression of the melting polymer and consequent removal of trapped air which is forced back through the feed section. This ensures an extrudate that is free from air bubbles. The function of the final (metering) section of the screw is to meter the molten polymer through the die at a steady rate and to iron out any pulsations.

### 9.1.2 Back pressure

Just in front of the die there is fitted a breaker plate supporting a screen pack which consists of a number of fine or coarse mesh gauzes. This has the effect of increasing the back pressure while also filtering out any contamination that might be present in the raw material. For any particular screw speed, a higher back pressure improves the mixing and homogenisation (and hence extrusion quality) but slightly reduces extruder output.

### 9.1.3 Temperature control

One of the most important factors governing the quality of the extrudate is control of temperature. Heat is supplied to the polymer raw material in two ways. The first is by external heating, by means of steam, oil or electric heaters while the second is by frictional heat which is generated internally by the shearing and compressing action of the screw. This frictional heat is quite appreciable and in modern, high-speed machines it is capable of supplying the whole of the heating required for steady running, with external heating only being necessary to prevent the extruder from stalling at the start of the run when the material is cold. The screw is also usually cored for steam heating or water cooling. Cooling is applied where the maximum amount of compounding is required. Again, this improves the quality of the extrudate at the expense of a slight reduction in output.

## 9.2 Sheet extrusion

As mentioned earlier, sheet extrusion is carried out by using a slit die. After emerging from the die, the sheet is led through heated rollers which are highly polished to give a good surface to the semi-molten sheet. This final polishing cannot, however, be used to cover up defects in the sheet caused by faulty extrusion and great attention is paid to eliminating surface defects before the sheet leaves the die. Possible faults include bad gloss (caused by too low a roll temperature or by defective rolls), continuous lines in the direction of extrusion (caused by die contamination or scored rolls) and continuous lines across the sheet (caused by too low a melt temperature, leading to jerky operation).

## 9.3 Extrusion blow moulding

In extrusion blow moulding the time taken for the blow moulding to cool is the major factor controlling cycle time. The mass temperature is usually kept as low as possible, therefore, subject to the production of bottles having good appearance and performance. Good homogeneity of the melt can only be attained, therefore, by the use of high back pressures and high shearing forces. This necessitates the use of robust extruders with good thrust bearings. Screws with high $l/d$ ratios (around 20:1) are preferred, with a short compression stage and a fairly long metering stage.

## 9.4 Coextrusion

An important development is coextrusion, whereby two or more extruders are coupled to a single die head to produce a multilayered parison or sheet with improved barrier or mechanical properties. The success of the technique depends very largely on the design of the die and the way in which the individually extruded melts are brought together prior to their extrusion as a multilayer. The production of multilayer bottles by coextrusion is dealt with in Chapter 11 but it should also be noted that multilayer sheet can be thermoformed to give multilayer tubs, trays, etc.

CHAPTER 10

## Injection moulding

---

## 10.1 Outline of process

The injection moulding process is a versatile one and consists essentially of softening plastics granules in a heated cylinder and injecting the melt under high pressure into a relatively cold mould where it hardens. The moulded article is then ejected from the mould by means of stripper plates or by compressed air. Injection moulding can be used for the production of articles ranging in weight from a few grams to several kilograms. The process is cyclical, the sequence of events being as follows:

1. The mould is closed and a locking force is applied.
2. A motor rotates the screw. The resultant shearing action on the plastics material, coupled with the conductive heat from the external heaters, melts the granules.
3. The molten material passes from the screw flights, through a screw tip non-return valve as the screw continues rotating.
4. The build-up of molten material in front of the screw causes it to move axially backwards.
5. When the required volume of molten plastic has accumulated at the front of the screw, the screw stops rotating and moves forward under pressure to inject the melt into the mould. The melt then flows through the 'sprue' opening in the die into 'runners' which terminate in 'gates' leading to the mould cavity itself.
6. Pressure is maintained on the screw for the period during which the material in the mould cools and contracts. This period in usually known as the 'injection time'.
7. The screw is retracted, but the mould remains locked for a period known as the 'freeze time'.
8. The mould is opened and the moulding is removed.

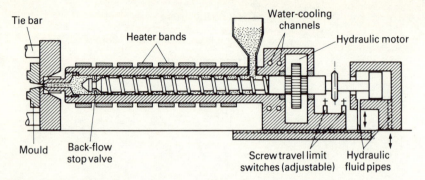

**Figure 10.1**  Single-screw injection moulding machine (with acknowledgements to ICI plc).

The plasticising unit of a single-screw injection moulding machine is shown in Fig. 10.1.

Expanding on the above, the gate is the point of entry into the mould cavity and may take various forms according to the viscosity of the melt and the size and shape of the mould cavity. Small gates are preferred in order to leave as small a scar as possible and to provide a smooth flow, but in general the thicker the moulding section, the larger the size of the gate.

Runners are channels which feed the molten material to the mould in cases where there is more than one gate. Runners connect at the other end with the sprue which is the path from the external entry of the material into the die, to the gate (if there is one only) or to the runners if there is more than one gate.

More than one gate may be necessary in a large mould, in order to equalise the flow of material through the mould cavity and so reduce the possibility of warpage when the moulding is cooled. When large numbers of small mouldings are to be produced it will usually be more economical to use a multi-cavity mould, so that a number of mouldings are produced during each moulding cycle. Each cavity will have its own gate so that a system of runners is again necessary to carry the molten material from the sprue to the various gates.

The positioning of gates is also important and it is usually advisable, for ease of cavity filling and evenness of flow, to feed into the thickest section of the cavity and in such a way that the material impinges on the core or the cavity wall.

## 10.2  Machine characteristics

Injection moulding machines are characterised by their shot capacity, plasticising capacity, screw diameter, injection speed, injection pressure and clamp pressure (mould locking force).

### 10.2.1 Shot capacity

The shot capacity may be given as the maximum weight that can be injected per shot, in which case it is usually quoted in terms of grams (or kilograms) of polystyrene or cellulose acetate. This has the disadvantage that because the figure is dependent on the swept volume of the cylinder during one stroke of the screw plunger and on the volumetric capacity of the feed mechanism, it is necessary to correct for bulk density and specific gravity when dealing with other materials. A better way of expressing shot capacity is in terms of the volume of material that can be injected by the plunger into a mould at a specific pressure.

### 10.2.2 Plasticising capacity

The plasticising capacity is usually defined as the number of kilograms per hour, of a particular material, that can be brought to the temperature required for moulding. Plasticisation is sometimes achieved by two-stage processes with separate components for plasticising (melting) and injection.

### 10.2.3 Screw diameter

A manufacturer may often supply a number of screws of differing diameters for use with a single injection unit. As they all have a similar maximum stroke length, the larger the diameter, the larger the swept volume. However, because the total maximum thrust is constant, the maximum injection pressure available is reduced.

### 10.2.4 Screw zones

The screw of an injection moulding machine is usually divided into three zones, each of which has a separate function. The first is the feed section which transports the plastics material from the hopper to the heated portion of the barrel. The flight volume is constant throughout this section. The second section is the compression section. Here, the material is compacted from its powder or granular state to that of a homogeneous melt. The volumes of the flights in this section decrease in the direction of the nozzle in order to compensate for the change of density of the material that occurs during this process. The geometry and position of the compression zone varies according to the melting or softening points of the material being moulded. An abrupt compression is necessary for nylon, which has a sharp melting point, while for PVC, which has no true melting point, the compression increases gradually along the length of the screw. Polymers such as polyethylene have a screw intermediate in design.

The third section is the metering zone where the final mixing and heating of the material into a homogeneous melt is carried out. This section also has a constant flight volume.

### 10.2.5 *Injection pressure*

The injection pressure of an injection moulding machine is the pressure exerted by the face of the plunger and can vary from about 70 MN/m$^2$ (10 000 lb/in$^2$) to 175 MN/m$^2$ (25 000 lb/in$^2$). Pressure losses in the system mean that pressures in the mould cavities are much less than the plunger pressure. Materials vary as to their pressure requirements while pressure requirements are also very different if the product to be moulded is a thin-wall container or a crate.

### 10.2.6 *Clamp pressure*

Notwithstanding any pressure losses in the system, the actual force exerted on the inside of the mould is still large, particularly if the projected area of the mould is large. The clamp pressure is, therefore, an important factor in determining the maximum projected area that can be moulded on a particular machine.

On large machines, the moulds are normally closed by means of large, direct-acting hydraulic rams while on the smaller models a 'toggle' action is often preferred. A toggle lock system is one of mechanical links used between the moving platen and the motive force. This motive force is also obtained from a hydraulic ram but one which is much smaller in size than would be required by a direct hydraulic machine of the same capacity. Combinations of toggle and direct hydraulic systems are also used.

### 10.2.7 *Injection speed*

This can be a very important factor in determining the output of an injection moulding machine. It is usually expressed as the volume of material that is discharged per second through the nozzle during a normal injection stroke. It depends on a number of factors including pressure, temperature, the material used and the size of the smallest aperture in the flow line. With low-viscosity materials and the use of fairly large gates, the injection plunger speed itself may be the limiting factor.

## 10.3 Machine types

It can be seen from the foregoing that the range of injection moulding machines is an extremely wide one. Apart from the different sizes, ranging from those with a shot capacity of a few grams to the giants capable of delivering several kilograms in one operation, there are reciprocating screw and two-stage units, automatic and semi-automatic types, those with hydraulic or toggle action locking and vertical or horizontal types. Vertical types are particularly useful where the moulding of inserts is required.

Development of the larger type of injection moulding machine has not been due solely to the demand for larger single mouldings. Because, for example, a 100 gram machine is slightly lower in cost than two 50 gram machines, there is a tendency to use larger machines with multi-cavity moulds. However, flexibility of operation is a point in favour of single-mould, faster cycling machines. Another factor to be considered is where several different-coloured mouldings are called for, and here again it is better to run several smaller machines rather than one large machine using multi-cavity moulds.

## 10.4 Machine variables

For any particular material, successful moulding is determined by mould design and by the correct setting of the following injection moulding machine variables.

### 10.4.1 Injection pressure

Injection pressures can vary quite widely and are dependent on mould design and on the size of machine. In general, the aim is the use of minimum pressure to produce full shot mouldings, free from such defects as surface sink marks or voids. Excess pressure should be avoided as this can lead to 'flashing', i.e. the escape of material from the mould joints. However, high injection pressures are vital for the production of thin-wall mouldings in order to achieve the necessary high injection rates. Molten resin moving through a thin cavity, over a long distance, tends to cool quickly and 'freeze-off' before the mould is completely filled.

### 10.4.2 Cylinder temperature

The object of heating the moulding material is to bring it to a suitably plasticised condition by the time it is ready for injection into the mould cavity. The temperature of the material depends not only on the temperature of the cylinder (or barrel) but also on the shearing action of the screw. Thin-wall mouldings, such as yoghurt pots, require higher cylinder temperatures than those with thick sections. This is because the thin-wall mouldings require less time to 'freeze' in the mould, thus reducing the overall time cycle and, consequently, the residence time of the moulding material in the heating cylinder.

### 10.4.3  Venting

Some plastics materials may contain volatile contaminants which, if not removed before moulding, could cause moulding defects known as 'splash marks'. One method of removing these volatiles is by passing a warm current of air through the machine hopper. Another is to pre-dry the material in an oven. A further method is the use of a vented barrel, in conjunction with an appropriate screw, whereby volatiles are removed from the plastics melt before the material is injected into the mould. A typical vented barrel and screw are shown in Fig. 10.2.

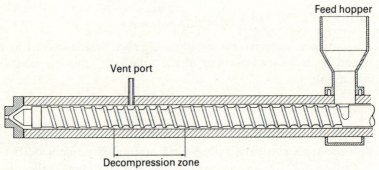

**Figure 10.2**  Typical vented barrel and screw (with acknowledgements to ICI plc).

The increase in flight depth of the screw beneath the vent relieves the pressure on the melt and allows volatile materials to vaporise and escape through the vent. Vented barrels and screws are of most use with hygroscopic plastics such as acrylics, polycarbonate and, particularly, nylon.

### 10.4.4  Cycle time

From the standpoint of economics, the aim should be the minimum time for each part of the moulding cycle consistent with good quality moulding. The speed of the injection stroke is governed mainly by the viscosity of the material (and hence by the melt temperature), the injection pressure and the minimum aperture size in the flow line. The minimum screw forward time is governed by the fact that if pressure is removed too early from the moulding, then surface 'sinking' or internal voiding can occur and it is, therefore, essential that the gate should freeze-off completely before the screw is retracted.

### 10.4.5 *Mould temperature*

Constant mould temperature, below the softening point of the material, is the aim and this is usually carried out by circulating constant-temperature fluid through channels in the mould. The mould temperature chosen is that which enables good mouldings to be produced with a minimum time cycle.

## 10.5 Mould design

Good mould design is a prerequisite for the production of good quality mouldings and no amount of juggling with machine controls can compensate for a badly designed mould. The subject of mould design is a highly complex one but some of the more general points are considered briefly in the next few paragraphs.

### 10.5.1 *Number of mould cavities*

There can be no single answer to the question of the optimum number of mould cavities, since it depends on the complexity of the moulding, the size and type of machine available, the cycle times likely from each possible number of cavities and the number of mouldings required. As far as the last factor is concerned, if a fairly accurate estimate of the various mould costs and cycle times can be made and a cost of running the machine (without material) is assumed, then a break-even quantity per unit of time can be calculated which can be compared with the total production required.

Cycle times are not increased pro rata with the number of cavities. This is because in a single-cavity mould the limiting factor is the cooling time of the moulding, whereas with higher numbers the plasticising capacity of the machine tends to be the controlling factor.

Another way of increasing the number of mouldings per cycle is by the use of stack moulds where two identical sets of cores and cavities are stacked together, back-to-back. Use of a stack mould can almost double the productivity of a machine and result in approximately 20% less energy consumption per moulding produced. Because of this, stack moulds are becoming increasingly popular.

### 10.5.2 Mould heating or cooling

As mentioned earlier, the control of mould temperature is usually achieved by circulating a constant temperature fluid through strategically placed channels. Initial mould design should make provision for these channels, bearing in mind such factors as symmetry of cooling and the effect of the temperature gradient across the mould on the flow of material. Cooling should be greatest at the mould gate, because the plastic that reaches the furthest extremities has more time to become cooled. This evens up the cooling times and so lessens the possibilities of warpage.

With certain materials, e.g. acetals and nylon, the mould is heated in order to achieve the desired degree of crystallinity and hence optimise the properties of the moulded part.

### 10.5.3 Mould venting

If no provision is made for venting the mould, then the air in the mould cavity will be trapped in some corner and then highly compressed. If this compression is rapid enough, the resultant rise in the temperature of the air may be high enough to cause scorching of the moulding at that particular point. With fairly simple shapes it will often be possible to arrange the mould in such a way that the point of venting is along the mould parting line. If such a solution is not possible then the appropriate place for venting can be found by trial and error. This is done by making trial shots and observing the plastic flow. Vents can then be placed in the mould where they give the most effective release of any trapped air. Suitable venting can usually be achieved by a slot about 6 mm (0.25 in.) wide and 0.05 mm (0.002 in.) deep. This will allow the release of air but is not large enough to allow the high viscosity plastic to escape.

### 10.5.4 Shrinkage of mouldings

One of the factors of good mould design is an allowance for shrinkage of the plastic. The amount of shrinkage varies with the material, being between 0.002 and 0.008 in./in. for polystyrene and between 0.015 and 0.025 in./in. for nylon 6,6 and polypropylene. Polyethylene poses particular problems because the rate of shrinkage is different in the direction of flow and across the flow of the material. This can lead to warpage and if a large, flat area is to be moulded it is best to have a number of gates in order to equalise the flow.

### 10.5.5 Gating of the mould

Gating of the mould has a very big effect on the quality of the moulding obtained and one example of this was briefly mentioned in the previous paragraph. In general, the aim is to produce the best possible flow conditions consistent with ease of finishing, while interfering as little as possible with the final appearance of the moulding. The type and size of the gate will depend on the type of moulding and on the material being moulded. As mentioned earlier, small gates are used for preference in order to leave as small a scar as possible and to provide a smooth flow but, in general, the thicker the section of the moulding, the larger the size of the gate.

Two common types of gate are illustrated in Fig. 10.3. The sprue gate is normally used where the cavity is difficult to fill or when feeding the runners of multiple cavities. Among the disadvantages of this gate are the length of time taken for it to freeze, the liability of surface sink marks opposite the sprue and the expense of trimming. The advantage of the sprue gate is that the pressure loss across the gate is low, and it is possible to utilise 'after pressure'.

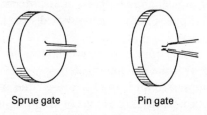

Sprue gate          Pin gate

**Figure 10.3**   Typical gates (with acknowledgements to ICI plc).

The pin gate, also pictured in Fig. 10.3 is preferable when cavities are easy to fill or when easy moulding materials are used. Advantages include ease of trimming, freedom from sink marks opposite the gate, a short freeze time and a tendency to homogenise the melt by frictional heat. Disadvantages are a high pressure loss and an inability to utilise after presssure.

### 10.5.6 Hot runners

In normal multi-cavity moulding the sprues and runners are cooled and removed at each cycle. This scrap material is then reground and remoulded. The use of cold runners can appreciably slow the speed at which the mould operates because the runner must be cooled prior to ejection. Also, the amount of material needed to fill the sprue and runners at each cycle is all part of the shot capacity of the machine and this reduces the possible size of the moulding. For short-cycle moulding or for articles of thin section (such as many packaging applications), the hot runner technique has become popular. A hot runner system maintains and controls the temperature of the molten material right up to the gates by means of electrical heaters. Another technique is that of insulated runners, which is often used with polyethylene and polypropylene. Here, the runners are of large diameter so that the outer layer of polymer freezes while the centre remains fluid. The frozen polymer then acts as an effective insulant between the 'hot runner' and the cooled mould. To ensure consistent quality, a hot runner system should be balanced so that it supplies molten material to each cavity at the same temperature and pressure. Flow balance is achieved by ensuring that each flow channel has an equal length and an equal number of turns while temperature balance is achieved by uniform heating and insulation.

### 10.5.7 Moulding removal

There are two main techniques for the removal of the moulding from the mould, namely, mechanical and compressed air. Mechanical methods are normally based on stripper plates which physically push the mouldings off the mould core. The stripper plate can be activated by hydraulic cylinders attached to the machine platen by air cylinders in the mould or by mechanical linkage with the machine platen.

Alternatively, compressed air is used to loosen and blow the mouldings off the core. Air ejection has the advantage of involving fewer moving parts and so less maintenance is required. The mould can also be more compact.

The ease of removal of a moulding is also a matter of design of the moulded piece. It is advisable, for example, to avoid undercuts and re-entrant curves, which restrict or prevent separation of the mould, while for deep mouldings it is necessary to provide a slight taper on the walls to facilitate extraction from the mould.

## 10.6 Quality of mouldings

The object of the injection moulder is to produce mouldings that fulfil the following requirements:

1 They are free from faults, such as voids, sink marks, etc.
2. They fill the mould cavity exactly.
3. They are easy to extract from the mould.
4. Their temperature on extraction is such that they will not subsequently distort.

Injection moulding quality, as judged by the foregoing criteria, is determined to a great extent by the pressure and temperature of the material in the mould cavity at the instant that the gate freezes. At that instant the mould cavity is exactly filled with the hot material under pressure. From this point on, the temperature drops with two related and opposing consequences. Normal thermal contraction tends to reduce the volume of the moulding but the pressure is also relaxed and this relaxation tends to force the material to expand. The two effects occur at the same time and exactly neutralise each other with the result that the volume of the moulding remains constant. This balance is maintained until constant temperature is reached or the pressure falls to zero.

In the ideal state, the pressure will have fallen to zero at the same time as the temperature has reached the correct value for extraction (i.e. the effects of thermal contraction have been counteracted exactly and the volume of the moulding has remained constant). This gives a full moulding that can be extracted easily.

If there is an appreciable residual pressure in the moulding after the effects of thermal contraction have been fully compensated for by pressure relaxation, then 'sticking' can occur with consequent difficulty in the extraction of the moulding. This is caused by the residual pressure attempting to relax and so increase the volume of the moulding.

Another possibility is that the pressure in the material falls to zero before the temperature reaches the correct extraction temperature. Since the pressure can fall no further, subsequent cooling to the extraction temperature leads to uncompensated contraction. This, in turn, leads to surface sinking or voids.

## 10.7 Moulding defects

Because of the interdependence of the various machine variables there is unlikely to be only one cause for any moulding defect. In addition, as mentioned earlier, many of the possible faults can be aggravated by bad mould design. Table 10.1 gives some of the possible ways of remedying them. Where several different remedies are available for one defect the optimum one must be found by trial and error.

## 10.8 Safety

The temperatures and pressures involved in injection moulding make safety an important factor in machine design. Front and rear gates are made to interlock mechanically, electrically and hydraulically in order to prevent clamp movement should the gates be left open or be opened

*Table 10.1* Some injection moulding faults.

| Name of fault | Possible cause | Suggested remedy |
| --- | --- | --- |
| Short mouldings | Insufficient material | Adjust feed setting |
| | Inadequate flow | Increase pressure |
| | | Increase temperature |
| | | Increase size of gate |
| | | Increase injection rate |
| | Unbalanced cavity in a multi-cavity mould | Check sizes of cavities |
| Flashing at mould parting lines | Insufficient locking force | Increase locking force |
| | Injection pressure too high | Reduce injection pressure |
| | Material too hot | Reduce barrel temperature |
| | Mould faces out of line | Re-bed mould faces |
| | Mould faces contaminated | Clean mould faces |
| Surface sink marks | Material too hot when gate freezes | Reduce barrel temperature or enlarge gate (to delay freezing) |
| | Insufficient dwell time | Increase dwell time |
| | Insufficient pressure | Increase pressure |
| Voids | Condensation of moisture on polymer granules | Pre-dry granules |
| | Condensation of moisture on mould surface | Increase mould temperature |
| | Internal shrinkage after case-hardening of outer layer | Increase pressure |
| | | Increase mould temperature |
| | | Enlarge gate |
| | | Increase dwell time |
| Burn marks | Air trapped in cavities | Improve mould venting |
| Weld lines | Material too cold | Increase barrel temperature |
| | Mould too cold | Increase mould temperature |
| | Injection pressure too low | Increase injection pressure |
| | Gates wrongly located (including too big a distance from gate to weld joint) or designed | Relocate gates and/or redesign |
| Distortion of mouldings | Ejection of moulding at too high a temperature | Reduce mould temperature |
| | High residual stresses | Reduce barrel temperature |
| | | Increase melt temperature |
| | | Reduce dwell time |
| | | Reduce injection pressure |

during the operating cycle. In most machines a low-pressure safety condition operates during clamp closing. During this part of the cycle, if any obstruction is encountered, the slight increase in force above that normally needed to close the mould signals the clamp to stop.

## 10.9 Automation and controls

The injection moulding process is capable of being highly automated, from the arrival of the polymer at the moulder's company to the packaging of the finished mouldings. At one end, polymer is delivered in road/rail tank cars, from where it can be pumped into large silos adjacent to the moulding plant. It is then pumped to colouring stations inside the plant where blending is performed automatically as the material is fed to the individual machine hoppers. After ejection of the mouldings from the machine it is often possible for them to be discharged on to conveyors, then automatically inspected, counted and packaged.

In between, there are such items as fast changing of moulds, robots for the high speed removal of mouldings and automatic quality control monitoring. Computerised controls are also widely available, including keyboard setting of the various machine variables and solid-state memory storage of optimised settings.

## 10.10 Structural foam moulding

As we have already seen, the normal injection moulding process involves very high injection pressures, hence moulds must be strong and high clamp pressures must be used. In structural foam moulding the injection moulding process is modified to produce an article having a foamed interior, integral with substantially unfoamed inner and outer skins.

The production of structural foam mouldings is based on the dispersal of an inert gas throughout the molten polymer, either by the introduction of gas directly into the melt or by premixing the polymer granules with a chemical blowing agent. When such a blowing agent is heated during processing, it decomposes and releases an inert gas (usually nitrogen). The pressure inside the injection machine barrel is high enough to prevent any foaming at that stage. A metered quantity of the polymer/gas mixture is then injected into the mould cavity where it expands and fills the mould. The foam tends to collapse at the comparatively cold mould walls, forming a tough skin. The collapse of the foam gives an attractive swirling finish at the surface of the moulding.

Because the mould is filled by expansion of the polymer/gas melt, external pressures can be low. This means that moulds can be less massive and hence cheaper, and smaller runs are economical. Additional advantages include the following:

1. A high stiffness to weight ratio exists, due to the foam-sandwich structure.
2. Moulded parts exhibit a lower degree of warpage and are basically stress-free because filling of the mould is achieved by foaming of the molten polymer and not by high pressures.
3. Sink marks are greatly reduced, thus allowing the moulding-in of ribs, bosses and intersecting walls. This, together with the possibility of large platen areas allows the injection moulding of large parts.

One limitation of structural foam moulding is the fact that the minimum wall thickness attainable is fairly high (around 5 mm). This is because below this figure the wall consists essentially of the two solid skins with a negligible foam content.

The structural foam technique is applicable to almost any thermoplastic that can be moulded by standard techniques. The main material usage is polystyrene, ABS, HDPE and polypropylene. Applications for structural foams include business machine housings, cutlery and paint-brush handles and many furniture items but the main packaging uses are for pallets and tote boxes.

CHAPTER 11

# Blow moulding

Blow moulding is a versatile process, designed to produce hollow objects, the basic techniques of which were developed by the glass industry. The principle of the process is that air (or sometimes nitrogen) is forced under pressure into a sealed molten mass of the plastic which is surrounded by a cooled split mould of the desired configuration. The air pressure causes the molten mass to expand until it reaches the mould walls where it is cooled by contact and is resolidified. The split mould is then opened and the bottle is removed. The preformed plastic mass that is blown into a bottle, vial, drum, etc. is known as a 'parison' (a term derived from the glass blowing industry). A parison is generally of tubular form and can be either injection moulded or extruded. The two main methods of bottle blowing, based on the methods of producing the parison, are referred to as injection blow moulding and extrusion blow moulding respectively. A third process was developed during the 1970s and is known as stretch blow moulding.

## 11.1 Injection blow moulding

Injection moulding of the parison was one of the earliest methods to be developed, probably because it most closely resembles the methods employed in the glass bottle industry. It is not an exact replica of the glass blowing process. The latter uses a solid 'gob' of molten glass as the starting point and this is then blown or pressed into the parison mould. The melt characteristics of glass and plastics are very different and early attempts to use solid 'gobs' of plastic in bottle production failed. However, a variation on the glass bottle parison pressing method has since been developed and is used to some extent for small containers. The process is known as displacement blow moulding. A pre-measured amount of plastic is placed in the bottom of the parison mould cavity through a nozzle which is moved upwards as the plastic mass is discharged. The resultant mass is, therefore, free from voids or air bubbles. The core of

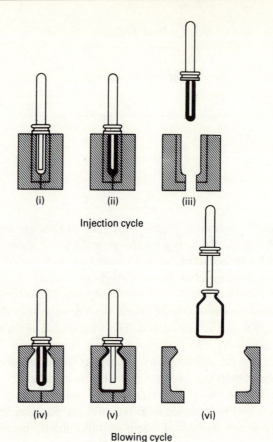

**Figure 11.1** Basic sequence of injection blow moulding. (Source: Miles and Briston, *Polymer Technology*, 2nd edition, 1979, Chemical Publishing, New York.)

the mould is then pushed into the molten mass which is displaced up and around the length of the core and into the neck finish area. The parison so formed is blown to its final shape in a second mould. The process is said to provide some of the advantages of injection blow moulding but with less moulded-in stress.

  The basic sequence for injection blow moulding is shown in Fig. 11.1. The molten plastic is injected into a parison mould around a core rod, or blowing stick. The cavity of the parison mould is accurately machined to determined dimensions, so as to give the required thickness profile, after blowing, to form the finished bottle. While still in the molten state, the parison (and its associated blowing stick) is transferred to a second mould, known as the blowing mould. Here, the final bottle shape is obtained by passing compressed air down the blowing stick and into the parison. The blown bottle is cooled by the walls of the mould, the mould is opened and the bottle is ejected.

Originally, the parison and blowing stick were transferred from the injection moulding machine to the blowing mould by hand. Automatic methods of transfer were soon developed, however, as were other methods of increasing production speeds. These include the mounting of blowing moulds on either side of each injection mould, with the parisons being transferred automatically into one or other of the blowing moulds alternatively. A further development is the use of multiple-mould cavities, thus permitting even greater increases in output speeds.

The chief disadvantage of the two-position methods mentioned earlier is that the injection mould and blow mould stations have to stand idle during removal of the moulding. This disadvantage has been overcome by means of the three-position method, invented by Gussoni in Italy, which uses a third station for bottle removal. Three blowing sticks are spaced radially at equal intervals about a pivot and in a plane between the faces of the mould. The moulds are so designed that an injection cavity and a blowing cavity are located at 120° to each other, to mate with two of the three parison sticks. The third stick is not enclosed by a mould but serves as an ejection station. In this system, the radially mounted parison sticks can be brought in turn to the moulding station, the blowing station and the ejection station.

There are a number of technical advantages associated with the injection blow moulding process. One is the accurate control of wall thickness that is possible. The blowing stick and parison mould can be pre-shaped to give walls of varying thickness so that there is a correct distribution of material after blowing into the final mould, thus avoiding thinning at base corners and at the shoulders. Secondly, injection blow moulding is also capable of producing a fully finished bottle. This is because the parison is completely enclosed and fully formed in the blowing mould. This latter is fully closed during blowing, whereas in most extrusion blow moulding systems the mould itself seals the parison at the bottom of the bottle. The excess material, known as 'flash', has to be removed from the bottle subsequent to its ejection from the mould. However, methods of automatic deflashing have been developed that do not interfere with the normal bottle cycle.

The injection blow moulding system is also more versatile in terms of the number of different plastics that can be used. Polystyrene, for example, is not easily blown on extrusion blow moulding equipment whereas perfectly good results can be obtained by the injection blow moulding process. Other advantages of the injection blow moulding system are that it is particularly suited to the production of very small bottles and that, in general, the surface finish is superior to that obtained by the extrusion blowing process.

However, there are also disadvantages to the injection blow moulding process. Equipment is usually more expensive (compared with that required for extrusion blow moulding) and the process tends to be slower because the injection moulding part of the cycle necessitates higher mass temperatures and so the cooling cycle is longer. One reason for the higher cost of injection blow moulding is the fact that two moulds are necessary as opposed to the one for extrusion blow moulding.

The injection parison mould normally consists of two mould halves for forming the body, a blowing stick, a base and a neck ring. The second (or blowing mould) is similar to an extrusion blowing mould, having two mould halves with an interior cavity corresponding to the outside profile of the finished bottle. The blowing stick performs three separate functions. One is to act as a core around which the parison is injection moulded. The second is to be a vehicle for carrying the parison from the injection mould to the blowing mould, while the third is to provide a channel to admit compressed air to the inside of the parison to expand it to size. The neck of the bottle is produced in its final form at the injection moulding stage and is, therefore, dimensionally accurate.

Injection blow moulding is used to a great extent for pharmaceutical and cosmetics bottles because the bottles are often small and precise neck finishes are usually important. The materials most commonly used are polypropylene, HDPE and polystyrene. In the case of polystyrene, injection blow moulding provides a degree of orientation which, in turn, increases the impact resistance. PVC was not injection blow moulded until the late 1970s because earlier grades were unable to stand up to the temperatures involved without degradation. Development of more stream-lined equipment has also helped, by cutting down residence time in the moulder. PET can also be injection blow moulded.

## 11.2 Extrusion blow moulding

In this method, the parison is in the form of an extruded tube that can be produced either continuously or intermittently. The former is the most widely used as it is the most suitable for the high-speed production of small to medium sized bottles.

### 11.2.1 Continuous extrusion

In the continuous system, a predetermined length of tube is trapped, at intervals, between the two halves of a split mould. Both ends of the tube are sealed by the closing mould and the trapped portion is inflated by compressed air, introduced via a blowing pin or needle. One disadvantage

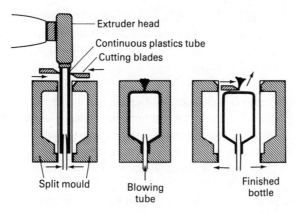

**Figure 11.2** Extrusion blow moulding process. (Source: Briston and Katan, *Plastics in Contact with Food*, 1974, Food Trade Press, London.)

of this method of inflation is that a finishing operation is necessary to produce an acceptable neck finish because the parison is gripped at both ends (by the two mould halves).

An alternative method introduces the compressed air via a mandrel inserted into the bore of the molten tubular parison. Because the hollow blowing mandrel has to enter the open end of the parison, there has to be a cutting device independent of the mould pinch-off to sever the parison and so provide the open end. One such method, where the extruded parison drops over a blowing tube at the base of the mould, is shown in Fig. 11.2.

Blowing can also be effected by the use of a mandrel inserted into the top of the mould. The bottle is then positioned neck-up and the internal diameter of the neck is determined by the external diameter of the mandrel. The cutting action between the mandrel and the top of the mould removes excess material and so produces a finished bottle. The cutting action is not always perfect (particularly as mould wear occurs) and a patented system, developed by the German firm Bekum, incorporates a sliding sleeve at the mandrel. This descends, after blowing, to form the top rim of the bottle and cut off excess material.

Methods of increasing production speeds include the use of more than one mould, the filled mould being moved away while another is moved into position, ready to receive the next portion of extruded tube. The moulds can be reciprocating ones or can be mounted on a rotary table. Each of these methods can be further adapted by having multiple die heads attached to the extruder and feeding them to multiple-cavity dies.

The continuous extrusion process is the one best suited for PVC. As mentioned earlier, PVC can degrade rapidly if slightly overheated. The relatively slow, uninterrupted flow of material in continuous extrusion reduces the formation of 'hot spots' which could cause degradation.

### 11.2.2 *Intermittent extrusion*

In the intermittent extrusion process the parison is rapidly extruded after the bottle is removed from the mould and the mould clamping mechanism does not have to move to a blowing station. Blow moulding, cooling and bottle removal all take place under the extrusion head. The stop/start aspect of this method makes it suitable for polyolefins but not for heat-sensitive materials such as PVC.

For bottles between about 0.25 litre and 9.5 litre capacity, the reciprocating screw method is used. This is similar to the process described earlier, under the heading of injection moulding. After a parison has been extruded, the screw moves backward, accumulating molten plastic in front of the screw tip. When the moulded bottle has cooled, the mould is opened, the bottle removed and the screw moves quickly forward again. This pushes the accumulated molten plastic mass through the extrusion head and so forms the next parison. It is possible to extrude up to twelve parisons simultaneously by the use of multiple die heads attached to the extruder.

The normal blow extrusion method is not suitable for the production of large mouldings, such as drums. This is because the parison needed to produce, say, a 210 litre drum would sag and deform under its own weight before extrusion could be completed. Such sagging would make it almost impossible to produce a moulding with an even wall thickness. A process has been developed, therefore, to overcome the difficulty. This is known as the accumulator-head process and consists of an extruder that feeds into a tubular reservoir which is an integral part of the extrusion head. A tubular plunger then rapidly extrudes the plastic melt from the head annulus using a low, uniform pressure that reduces the overall stresses.

One disadvantage of extrusion blow moulding in general is the fact that even a parison of constant wall thickness will not produce a bottle of regular wall thickness because the material is thinned more at the mould extremities than elsewhere. The base corners and shoulders of the finished bottles are weaker, therefore, than is desirable. One means of achieving thickness control is the so-called dancing mandrel. This is a specially designed conical mandrel which is raised or lowered within a mating conical section of the die ring to give a controlled variation in the clearance between the two. The thickness control is programmed to operate with each parison production cycle.

Extrusion blow moulding is used to produce bottles for a large number of products including bleach, liquid detergents, milk, shampoos, etc., and

bottling plants for blowing and filling. Production is more efficient since a minor breakdown in one of the stages does not cause a shutdown in the other. Process parameters can also be optimised. As an example, limits on parison production will not force a compromise in parison design to achieve higher production.

Another advantage claimed for the two-stage approach is that it permits more frequent quality control checks. Faulty preforms can be identified immediately after moulding and one does not have to wait for the finished bottle to discover the defect. This is claimed to cut the reject rate considerably but on the other hand there are said to be many faults inherent in the preforms that can only be detected in the finished bottle.

## 11.4 Blow moulding developments

There have been a number of developments in the basic processes already described and two of these are dealt with in the following sections.

### 11.4.1 Multilayer blow moulding

The performance and marketing requirements of many end uses are sometimes so complex that no one material can be satisfactory except, perhaps, at an uneconomic thickness. For example, polyethylene and polypropylene are relatively low in cost, are excellent barriers to moisture vapour and are suitable for food contact uses. However, they are poor barriers to oxygen and hence are not suitable materials for the packaging of many oxygen-sensitive products requiring a long shelf-life. On the other hand, ethylene/vinyl alcohol copolymers (EVOH) are relatively expensive, high gas barrier materials but are subject to deterioration of these properties by water. A thin (low cost) layer of EVOH, sandwiched between two layers of polyethylene or polypropylene, can often provide the answer.

The technique whereby multilayer bottles can be manufactured is based on coextrusion, a process that consists of coupling two or more extruders to a single die head. A multilayer parison is produced which is then blown to give a multilayer bottle. A tight control of extrusion process variables such as melt pressure, screw speed and melt temperature is necessary. For example, the temperature of each polymer must be closely controlled in order to maintain the right level of viscosity in the die. The success of the technique depends very largely on the design of the die and the way in which the individually extruded melts are brought together, prior to their extrusion as a multilayer parison. Parison thickness programming is also more critical than in conventional extrusion blow moulding.

Coextrusion provides the facility for producing individual plies of very low gauge and so it is possible to produce bottles where a thin layer of a relatively expensive high barrier material is sandwiched between inner and outer layers of low-cost material. Disadvantages of coextrusion include the difficulty of using the scrap produced during extrusion but it is possible to use regrind material as a blended core material with polyamide (nylon) as an outer glossy scratch-resistant surface layer. Where adhesion of adjacent layers is a problem then adhesive (or tie) layers are used.

Equipment costs are higher than for standard blow moulding machines, one estimate being 40–60% more for equipment blowing containers of capacity up to 1 litre. On the other hand, coextrusion is claimed to be cheaper than stretch blow moulding. Coextruded bottles are also said to be competitive with glass and metal in areas where single-layer plastic bottles have, hitherto, been unsuccessful.

A combination of multilayer bottles and stretch blow moulding can also be achieved by using a coinjection moulded preform.

### 11.4.2 Blow/fill/seal systems

Machines have been developed that will blow a bottle, fill it and seal it while it is still in the mould. One of the first systems to be developed consisted of an extruder which produced a vertical parison. This was gripped between the two halves of a split mould and then transferred to a blowing station where a double-channel blowing/filling spigot was inserted into the parison from above. Compressed air was blown down one channel to inflate the parison, after which the product to be filled was passed down the second channel. After filling, the spigot was withdrawn and the bottle sealed by crimping the still-hot bottle neck. Filling speeds were low and later developments concentrated on making the process a continuous one. One such development is the Bottelpack machine which utilises a continuous multi-mould system, with the addition of multiple parison heads, according to the number of moulds to be used. The operation can be followed by reference to A–E in Fig. 11.3. The mould halves are mounted on converging caterpillar tracks and move vertically downward (A). At a point below the parison head of the extruder, the two chains of mould halves meet and trap the tubular parison (B). This is inflated with air and then filled, using a multiple channel spigot. After filling, the spigot is withdrawn and is relocated in the next mould which by this time has closed round the next parison (C). Meanwhile, the first mould moves to the sealing head where the bottle is heat sealed. Finally, the caterpillar tracks diverge, the mould opens (D) and the filled bottle is ejected (E).

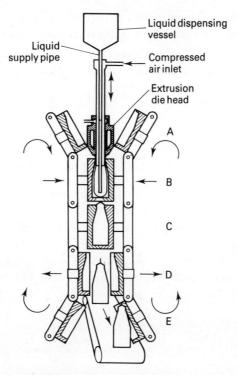

**Figure 11.3** Bottelpack 4000 machine. (Source: Briston and Katan, *Plastics in Contact with Food*, 1974, Food Trade Press, London.)

An important advantage of this system is that at no time does the liquid being filled come into contact with the atmosphere. This means that aseptic filling is possible because sterile materials filled into the bottles will not be contaminated by airborne bacteria. This is particularly important in the packaging of foods and pharmaceuticals. It is also possible to make bottles with very thin walls because they do not exist as unsupported (empty) bottles but remain in the mould until they have been filled and sealed. In normal bottle blowing the minimum bottle weight is often determined by the performance requirements on the filling line rather than by its performance in use. Such thin-walled bottles mean lower material costs.

Speeds of 3000 × 0.5-litre bottles or 1000 × 2-litre bottles per hour are claimed. Blowing cycles are usually shorter because some cooling of the bottle is effected by the product being filled. Conversely, the product is heated by the hot plastic but the temperature rise is small, as can be seen from the following example.

If we take the weight of a 1 litre bottle as 50 g and it is cooled from 180 to 50 °C by the product being filled then we can say that:

Weight of product × temperature rise of product × specific heat of product
= weight of bottle × temperature drop of bottle × specific heat of plastic

For an aqueous product and a low-density polyethylene bottle of 1 litre capacity we have (if $T$ is the temperature rise of the product):

$$1000 \times T \times 1 = 50(180 - 50) \times 0.5$$

$$T = \frac{50 \times 130 \times 0.5}{1000}$$

$$= 3.25\,°C$$

As the mould also contributes to the cooling of the bottle, the amount of heating of the product by the bottle is even less than the above.

Decoration of the filled bottles by silk screen printing is technically possible but the rejection rate can be quite high. Where printing of empty containers is concerned, the rejects can be granulated and the material reused but for filled bottles rejection may also involve loss of product. It is advisable, therefore, to carry out decoration by in-mould embossing or by labelling.

CHAPTER 12

# Thermoforming

In thermoforming, a heat-softened plastic sheet is formed either into or around a mould. An early example of this technique is 'bubble blowing' poly(methyl methacrylate) sheet to form aircraft canopies or domestic baths but there are now many variations both in basic techniques and in their degree of automation. In general, thermoforming techniques are best suited for producing mouldings of large area, for thin-walled mouldings or where only short runs (including the production of prototypes) are required.

## 12.1 Vacuum forming

The basics of the vacuum forming process can be seen by reference to Fig. 12.1. The equipment itself consists of a vacuum box, with an air outlet and a clamping frame, a mould, a heating panel and a vacuum pump. The mould, which is partly hollow underneath and is perforated, is placed over the air outlet. The thermoplastics sheet is then placed over the open top of the vacuum box and securely clamped by means of the frame, giving an airtight compartment. The sheet is heated until rubbery, the heater is withdrawn and the air in the box is rapidly evacuated by the vacuum pump. The sheet is forced down into close contact with the upper

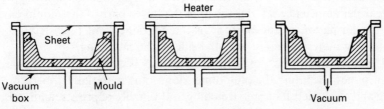

**Figure 12.1** Vacuum forming sequence. (Reproduced from Briston, *Plastics Films*, 3rd edition, 1989, Longman, London.)

surface of the mould by atmospheric pressure. Contact with the mould cools the sheet sufficiently for it to retain the moulded shape. The clamping frame is then released, the moulded article is removed and any excess sheet material is trimmed off.

Either male or female moulds may be employed, the simplest method being vacuum forming into a female mould. However, only simple designs are then possible because of excessive thinning at the bottom and corners of the mould cavity. With a male mould the thickest section is on top instead of the side walls. Variations on the basic vacuum forming technique have been developed which help to overcome some deficiencies of the basic method. One of these is drape forming, where the mould is mounted on a piston inside the vacuum box. The piston rises and pushes the mould into the heat-softened sheet just before the vacuum is applied. This gives a certain amount of pre-forming to the sheet and so lessens thinning of the sheet at the corners of the mould (see Fig. 12.2).

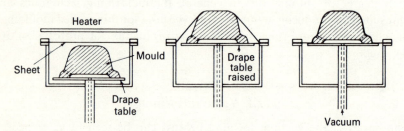

**Figure 12.2**  Drape forming sequence. (Reproduced from Briston, *Plastics Films*, 3rd edition, 1989, Longman, London.)

Plug assist is another variation which is particularly valuable for deep drawing, giving a more even wall thickness and lower cooling cycles. It consists of a hydraulic ram carrying a pre-form mould, profiled roughly like the final mould. This is pushed down into the top of the sheet, immediately prior to the application of the vacuum (see Fig. 12.3).

A heated 'air slip' around the plug is often employed to avoid marking of the sheet and to give a more even stretch. The function of the plug is to transfer sufficient material towards the lower part of the mould cavity for the formation of the bottom of the container and to distribute the remaining material evenly over the side of the moulding.

The final distribution of the wall thickness is dependent on a number of variables, including the plug dimensions, plug material, temperature of plug and sheet, depth of pre-stretching and velocity of pre-stretching. Plug material and plug temperature should be considered together. Good results can be obtained, for example, by using a plug made from

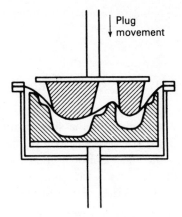

Plug
movement

**Figure 12.3** Plug assist preforming. (Reproduced from Briston, *Plastics Films*, 3rd edition, 1989, Longman, London.)

aluminium or some other heat-conducting material but the plug must then be heated within a sharply defined temperature range. If the temperature is too low, for example, then the high thermal conductivity becomes a disadvantage. The plug chills the sheet excessively and results in a container having a thick base and thin side walls.

Another important variable to be considered when using plug assist is 'slip-in' of sheet material into the mould. When the sheet is heated it expands and, if it is clamped only at the perimeter of the clamping frame, surplus material from the surrounding scrap can be drawn into the mould. This slip-in is irregular or uncontrolled because of the shape of the scrap areas, particularly in multi-cavity moulds. If, however, the sheet is clamped at the perimeter of the mould then no slip-in is possible. The third possibility is controlled slip-in. This possibility arises where the clamping edge forms an annulus around the mould aperture at a distance of about 3–5 mm. This amount of material is then available for controlled slip-in. Increasing the distance between individual moulds allows the annulus width to be increased, giving more slip-in and, consequently, a better wall thickness distribution.

A further variant on vacuum forming is bubble or air assist. In this process the sheet is pre-stretched into a bubble by pumping air into the vacuum box prior to the mould table rising. The possibility of thinning at the sharper or higher parts of male moulds in particular is again reduced. If too large a bubble is formed there is the danger of surplus sheet material when forming. This can be avoided by fitting a photo-electric cell and scanning beam at a suitable height above the sheet so that the air is automatically cut off when the bubble breaks the beam.

The air blowing feature can also be used for blowing the finished forming off the mould.

## 12.2  Pressure forming

Pressure forming is the same as vacuum forming with the exception that a positive air pressure is applied to the sheet from above, which again has the effect of forcing the softened sheet on to the mould. The pressure that can be applied to the sheet is not limited to atmospheric pressure which is, of course, the case with vacuum forming. Pressure forming, therefore, gives better reproduction of mould detail.

## 12.3  Matched mould forming

In this method the heated sheet is formed into shape by trapping it between matched male and female moulds. The mould detail, as one would expect, is even better using this technique but it is more expensive in tooling costs and the mould halves have to be made to tight tolerances.

## 12.4  Machine variables

There are a number of machine variables that can affect the quality of the finishing moulding. Among them are heating, cooling and mould design. Other factors include trimming and, sometimes, decoration.

### *12.4.1  Heating*

Various types of heat source have been used for heating the thermoplastics sheet prior to thermoforming, including contact heating, infra-red heating, high frequency heating and heat transfer using hot air or hot fluids. The first two are the ones usually used but whichever method is used, it should meet the following requirements. The sheet must be heated to its optimal forming temperature and this temperature should normally be even over the whole forming area. Temperature deviations between both sides of the sheet should also be kept to the minimum. The heating period should be as short as possible but excessive heating of the material must be avoided because of the risk of thermal degradation of the sheet. Energy consumption should also be kept as low as possible. It is desirable to reach the operating temperature as quickly as possible and so heater capacity should be adjustable.

Contact heating is particularly useful when the sealing flange area of the container has to be of uniform thickness for effective sealing in use. The sheet is clamped at the future sealing flange and is neither heated nor drawn out in this area. In contact heating the heating platen is often covered with a PTFE separating layer to prevent sticking of the sheet to the platen. If no PTFE layer is used, the heater platen surface is sometimes roughened by sand blasting to allow enclosed air to escape. Where double-sided heating is necessary (for thick sheet or for materials with low thermal conductivity, such as the polyolefins or expanded polystyrene) or when plug assist is being used, separate heating and forming stations are used.

Infra-red heaters are of two main types, namely, quartz or ceramic. Many of the larger machines have heating panels consisting of a large number of separate elements wired up in 'chessboard' fashion, thus allowing more heat to be switched to certain parts of the sheet if required. The heaters are positioned fairly closely to the sheet, about 100–150 mm (4–6 in.) away, in order to give efficient utilisation of the heating capacity. Double-sided heating is again available if required. Infra-red heating acts by absorption of the radiation by the sheet at the surface layers. The generated heat then flows by conduction through the rest of the material. Overheating of the surface can easily occur, while another possible consequence of too close an approach by the heaters is the impression of a heater pattern on the sheet.

### 12.4.2 Cooling

Cooling of the heated sheet will occur at the walls of the mould, which are often water cooled. All thermoplastics are poor conductors of heat and the mouldings should be removed from the mould as soon as they are dimensionally stable in order to reduce the cycle time. With thicker mouldings, cooling is often assisted by blasts of air or by water-mist sprays but this is not normally necessary with thin sheet. A certain amount of shrinkage takes place depending on the particular plastic used and on the intensity of cooling. Although shrinkage is basically undesirable it does facilitate the removal of mouldings from female moulds, especially those with vertical side walls.

### 12.4.3 Mould design

The construction of thermoforming moulds is governed by a number of factors. These include the sheet material to be formed and the surface finish required, the length of run (this affects the choice of manual, semi-automatic or automatic production), the machinability of the mould material and its thermal conductivity, the tolerances required and the allowable cost. For making prototypes or for small runs, moulds can be made from wood or plaster of Paris but for long-run production work epoxy resins or metals (usually aluminium) are used. The main advantage of epoxy resin moulds is their ease of manufacture and excellent machinability but their thermal conductivity is low. However, it is usually possible to cool the mould by circulating water within the mounting plate. Low thermal conductivity may sometimes be an advantage (for prototype work or small runs, for example) because major warming-up of the mould is unnecessary and the incidence of chill marks is greatly reduced.

Where the number of items to be produced justifies the extra cost then metal moulds should be used. Their main advantages are good mechanical resistance (long life), good dimensional stability, excellent thermal conductivity and the excellent surface finish that can be obtained on the moulded article. The disadvantages are the high cost of the material and the machining.

In general, male moulds are easier and cheaper to make than female moulds and they are usually used for deeper formings. The sides of the mould should be slightly tapered to facilitate removal of the finished moulding. With female moulds, ease of removal is aided by the shrinkage referred to earlier. One thing to be avoided when designing thermoforming moulds is the existence of deep and sharply angled corners as they cause severe thinning of the sheet. It is possible to mould undercuts but this usually necessitates the use of removable sections and so complicates the mould and lengthens the moulding cycle. The vacuum holes drilled through the mould must be kept small otherwise the sheet will be drawn into them.

### 12.4.4 Trimming

Trimming equipment can vary from a simple knife to a sophisticated cutting press. A roller press is often used for small to medium size mouldings and high production rates are possible. For larger mouldings a power-driven router together with a suitable cutting jig may be more effective. The separation of multi-impression mouldings can also be achieved using a combination of slitting knives with a cross-cut guillotine. The advantage of this method is that it is easily adjustable to accommodate a wide range of mould configurations but it can only be used where the separate mouldings have a rectangular shape. If a roller press is used, a separate cutting forme is required for each mould configuration. Roller presses can be used in-line or they can operate as separate units.

It should be noted that the use of a roller press enables the separated mouldings to be given rounded corners, whereas slitting and guillotining can only give square-cut corners. Rounded corners may be preferred from the marketing point of view and there is also less likelihood of corners being damaged.

### 12.4.5 Automatic operation

The thermoforming operation can range from manual to fully automatic. Automatic timers are available which will enable the complete cycle to be set in motion at the push of a single button, with the cycle including any or all the features such as drape, plug assist, etc.

A typical sequence of operations for a vacuum forming machine, once it has been loaded with sheet, is as follows:

1. push start button
2. heater brought forward for pre-set time
3. heater retracted
4. air injected to form bubble (where required)
5. drape table brought up or plug assist brought down
6. vacuum is drawn
7. air blast or water mist spray applied to surface of moulding
8. vacuum and air turned off; plug assist (if used) is withdrawn
9. air pressure applied to release forming from mould
10. air pressure off
11. drape table brought down or plug assist brought up
12. forming removed and machine reloaded with fresh sheet

For thinner sheet, the sheet feed can be replaced by continuous reel feed.

### 12.4.6 Decoration

Thermoformed articles can be decorated by most of the processes suitable for plastics generally. Thus, they can be labelled, printed, embossed, flocked or metallised. Printing can be carried out on the finished moulding or on the flat sheet, prior to forming. In the latter instance, the printed design has to be distorted in such a way that the appearance after forming is as required. Flexographic, screen or gravure printing can all be employed.

The degree of distortion necessary can be determined by printing a grid pattern or a system of concentric rings on the flat sheet. The amount of distortion that occurs at various points on the surface when the sheet is formed, gives an indication of the distortion to be allowed for. Another method entails printing the desired design on a previously formed article.

The printed moulding is then clamped in the thermoformer again and heated to a temperature above its softening point. The sheet returns to its original flat form and the amount of distortion necessary for the correct design can be seen. In either case it is important to use a homogeneous moulding, free from tension and with minimum thickness variation.

Very shallow formings can usually be printed without distortion as the sheet is subject to slight stretching only and this has very little influence on the final print. With high draw ratios the print design should be such that some slight distortion can be absorbed without affecting the overall appearance, because it is almost impossible to avoid all distortion under such conditions.

## 12.5  Solid-phase pressure forming (SPPF)

This process was developed mainly for polypropylene. Normal pressure forming is a well-established process for polystyrene, PVC, etc., but polypropylene is at a disadvantage in this area because the sheet is difficult to control at conventional pressure forming temperatures, and cooling times are longer than for polystyrene and PVC. The SPPF process effectively overcomes these processing difficulties by carrying out the forming at temperatures below the crystalline melting point of polypropylene. At such temperatures cooling times are reduced and overall cycle times similar to those of PVC and high-impact polystyrene can be attained. An additional advantage of the process is that the clarity of the mouldings is improved.

The sheet is heated in the normal way, the heat input being adjusted to bring the sheet to a uniform temperature of 155 °C. The heated sheet is first stretched into the mould cavity, using a heated plug. At the lowest point of travel, cold air is forced into the plug chamber under high pressure, thus pushing the sheet against the cooled inner wall of the mould and finishing the forming operation at high speed.

One other advantage of the lower forming temperature is that the polypropylene containers are free from the odour and taint that are sometimes present at normal forming temperatures. This odour/taint has often debarred polypropylene thermoformed containers from food-contact applications, in spite of their suitability from other points of view (their high softening point, for example).

### 12.5.1  Melt-to-mould rotary thermoforming

One reason for developing processes such as SPPF was to partially overcome the main disadvantage of thermoforming, i.e. the double heating involved (first, the heating of polymer granules during sheet extrusion; second, the reheating of the sheet during thermoforming).

Another answer is to thermoform the extruded sheet while it is still in the plastic state. Several attempts have been made, the latest of which is melt-to-mould rotary thermoforming. Here, the polymer is extruded directly from a slit die on to rotary moulds, giving a high productivity. Built-in thermal stresses are also reduced.

## 12.6 Blister packaging

A large outlet for thermoforming is that of blister packs. These consist essentially of a formed plastic sheet with a flange that is then fixed to a backing card by staples, adhesives or heat sealing. The blister may be a simple shape such as a circle or it may be contoured to fit the contents. Blisters may be single or multi-cavity, the latter being particularly useful for the packaging and display of sets of objects.

Blisters can be bought-in ready made and this could be the answer when simple shapes (or large numbers of more complex shapes) are required. On the other hand, the product manufacturer can set up his own thermoforming equipment and so become independent of outside suppliers. Blisters are usually formed on multi-impression moulds which are separated by slitting knives and a guillotine or by a roller press (as described in section 12.4.4).

If the blisters are to be heat sealed to the backing cards then the card is coated with a heat-seal adhesive. Filling is carried out by placing a number of blisters in a jig (with the flange uppermost); the object or objects are placed inside and the cards placed on top. The heated press is brought down and the sealed blisters are removed from the jig. Sealing speeds are increased by the use of rotary tables so that operations can be made continuous. Various degrees of automation are possible, including automatic loading of the heat-seal-coated boards into the sealing area and automatic pack discharge with the sealed packs being passed on for further handling to a work table. Loading of the blisters and the objects to be packaged is usually carried out manually but automatic loading is possible for suitable products of simple shape.

## 12.7 Thermoform/fill/seal (TFFS)

The thermoforming process is well adapted to form/fill/seal operations. In one type of machine, for example, blisters are formed to a profile roughly similar to the article to be packed. Articles are fed continuously into the blisters and a reel of fibreboard is sealed over them. The individual packages are then trimmed and separated. This type of operation is used for the packaging of high sales volume items such as razor blades, reels of cotton, etc.

The other main area of thermoform/fill/seal is the packaging of a wide range of foodstuffs, including jam, cream, sauces and cheese. Two reel-fed plastics webs are used, the first being formed into a series of tray-like depressions by heating and drawing a vacuum through the base of appropriately shaped moulds. The formed sheet is then indexed under a filling head and the filled compartments are lidded by sealing the second web of material on top. The web of filled and lidded containers is then cut and the individual packs separated. The web used for lidding is often preprinted. A thermoplastics-coated paper or aluminium foil may also be used for the lidding operation.

A somewhat similar technique is used for the packaging of tablets, pills or capsules into 'push-through' blister packs. A multi-cavity mould is used and the pills, etc., are fed automatically into the thermoformed cavities. The coated aluminium-foil lidding web is then brought down over the filled plastics web and heat sealed to it. When required, the individual pills, etc., are dispensed by pushing them from the plastics side through the relatively brittle aluminium foil.


## 12.8 Possible defects

Imperfect thermoformings can arise from various sources and may be associated with faults in the original sheet or with the thermoforming process. Imperfections can be divided into two main classes, namely, those which may affect performance and those which affect the appearance.


### 12.8.1 Imperfections affecting performance

Holes and cracks in a thermoformed moulding will obviously affect the ability of the blister, pot, tub, etc. to contain the product and must be classified as critical defects. Such defects should be rare if the mould has been correctly designed (see section 12.4.3) but occur in moulds of borderline design if other parameters are also out-of-line. Too high a speed of plug or drape table, for example, or too high a sheet temperature can also contribute.

Tests should be carried out if a change of sheet material is contemplated, perhaps to improve end-use performance. The mould design and production conditions suitable for one material may well be unsuitable for another.

Less obvious but still important is the existence of a thin spot (or spots) in the moulding. The minimum wall thickness required will vary according to the type of moulding and the end use envisaged but quality control checks must be carried out to ensure that imperfect mouldings are rejected.

Factors leading to thin spots are similar to those mentioned earlier for holes but to a lesser extent.

### 12.8.2 *Imperfections affecting appearance*

One possible imperfection, mentioned earlier, is the existence of a 'heater pattern' due to the heaters being too close to the sheet. Other imperfections such as 'fish-eyes' arise from faulty sheet material and must be prevented at source. Other visual defects may be due to scratches or embedded foreign matter or to 'drawing lines' on the side of the moulding, sometimes due to too low a forming temperature.

CHAPTER 13

# Rotational moulding

## 13.1 Processing

The rotational moulding process, as it exists today, has evolved from earlier techniques such as the Engel and Heissler processes. In the Engel process an open-ended mould is filled completely with a powdered plastic (usually LDPE) and a screw cap is fitted. Heat is applied to the outside surface of the mould so that the plastic immediately adjacent to the mould is fused, the heating being continued until the required wall thickness is built up. The mould cap is then removed, the mould is inverted and the excess plastic powder poured out. The inside of the moulding is then heated in order to fuse the inner surface, the mould is cooled and the moulding removed. This technique was used to produce large, open-ended mouldings with thick walls, where accuracy of wall thickness was not of paramount importance. One disadvantage of this process is the necessity of using bulk material which later has to be dumped out. This leads to the risk of contamination, either by semi-fused material or by foreign bodies.

The Heissler process was an attempt to overcome the disadvantages of a static process such as the Engels. A female mould is charged with a powdered plastic, the amount being governed by the required weight of the finished moulding. The mould is rotated on its horizontal axis while being simultaneously rocked on its vertical axis. At the same time, heat is applied externally to the mould, causing the powder to fuse. Heating is continued until fusion is complete; the mould is then cooled and the moulding removed. Although this process removes the necessity of using excess material it does not completely solve the problem of uneven wall thickness.

Rotational moulding is also a sinter (or fusion) casting process based on the use of plastics powder, rather than granules. The basic principles are as follows:

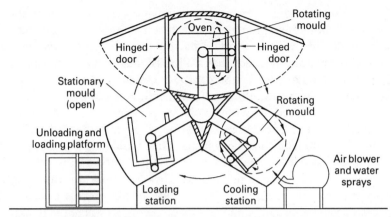

**Figure 13.1** Rotational casting. (Source: Briston and Miles, *Polymer Technology*, 2nd edition, 1979, Chemical Publishing, New York.)

1. A female mould is loaded with the exact weight of powder needed for the moulding and then closed.
2. The closed mould is then rotated about two axes, simultaneously, the two axes being perpendicular to each other. At the same time, heat is applied which causes the plastic powder to sinter (or fuse) over the whole internal surface of the mould in an even layer.
3. When all the powder has melted and is adhering to the inside surface of the mould the cooling cycle is commenced.
4. Rotation of the mould is continued until the moulding can be removed without distortion occurring.
5. Rotation is stopped and the moulding is removed.

A multistation machine is shown in Fig. 13.1.

Moulds for rotational moulding can be of light construction as there are no high pressures involved such as are present in injection moulding. The moulds are rotated at fairly low speeds, just sufficient to ensure good flow of the molten polymer. Because of the low pressures used, it is possible to produce mouldings that are virtually stress-free.

Cast aluminium, electro-formed copper–nickel and fabricated steel or aluminium have all been used as mould-making materials. Cast aluminium rotational moulds are much lower in cost than injection moulds, being comparable to moulds used in blow moulding or thermoforming. Fabricated moulds are also fairly inexpensive. The capital cost of rotational moulding equipment is also lower. In injection moulding, considerable pressure is required to keep the moulds closed during the moulding process and the same holds true, to a lesser extent, with blow moulding. In addition, high pressures are needed to inject or form the plastics material in these processes.

With a few exceptions, there are practically no shape or size limitations to rotational moulding although the finished item must be removable, i.e. with no undercuts, unless at mould parting lines. In general, as long as the mould configuration is such that it allows contact between the material and the inner mould surfaces and can be contained and rotated within a heated chamber so as to allow uniform heating of the mould, then the shape can be produced. Mouldings can be produced in a wide range of wall thicknesses but one design limitation is the fact that varying wall thicknesses are difficult to achieve in the same moulding. Some variation in wall thickness can be achieved by altering rotation ratios, insulating mould sections or by differential heating. However, thickness cannot alter sharply within a short distance. The nature of the process causes wall thicknesses to be unusually uniform except in corners where the thickness increases. This is an advantage inasmuch as this is where the greatest strength is required.

One process which eliminates the need for an enclosed oven is the jacketed mould machine. The moulds are made with a jacket around the outside surface, through which is pumped alternately hot oil for the heating cycle and cold oil for cooling. This leads to savings in floor space but mould costs are very much higher.

## 13.2  Stress cracking

One of the early problems with the rotational moulding of polyethylene was environmental stress cracking. As mentioned earlier (in Chapter 2), the tendency towards stress cracking is reduced by the use of high molecular weight (low melt index) polymers. Unfortunately, rotational moulding requires the use of relatively high melt index polymers in order to achieve satisfactory flow-out during the heating cycle. Polymers having a narrow molecular weight distribution give good results, with a reasonable compromise between stress crack resistance and flow-out. Stress cracking can also be reduced by proper attention to mould design, so as to reduce stresses and by avoiding the use of emulsion-type release agents. Such release agents contain surface active agents which are severe stress crack promoters.

## 13.3  Applications

One of the attractions of rotational moulding is that it is possible to produce sizes and shapes that were previously restricted by machinery and mould limitations. A plastic part can now be formed in almost any shape and with no upper limit on size (within the bounds of common sense!). Many applications, therefore, are in non-packaging fields such as automotive (petrol tanks in nylon), toys (PVC, low- or high-density polyethylene, nylon, etc.) and water tanks. However, many packaging uses do exist, including plastic drums, rectangular containers, transit trays and intermediate bulk containers.

# CHAPTER 14

# Moulding and extrusion of expanded polystyrene

## 14.1 Moulding

As mentioned in Chapter 3, polystyrene is available in an expandable form which can be processed by steam moulding to give lightweight articles with many outlets in the field of packaging. There are two basic methods for the production of expandable beads, namely, by polymerising and gassing the beads in one step, or by polymerising the styrene, then impregnating the polystyrene beads at elevated temperature and pressure. The gassing agent in both cases is normally pentane although other agents have also been used.

Moulding of expandable polystyrene beads is carried out in three stages: pre-expansion, maturing and moulding. Pre-expansion is achieved by heating the expandable beads in steam. This has the dual effect of increasing the pressure of the blowing agent within the beads and of softening the polystyrene. The beads thus expand, the degree of expansion (and hence the final density) being controlled by the temperature and the duration of heating. Pre-expansion can be carried out as a batch operation or continuously.

The pre-expanded bead consists of non-communicating cells of 80–150 µm diameter with about 550 000 cells/cm³. The bulk density of the original expandable polystyrene beads is about 673 g/litre (42 lb/ft³) and expanded articles can be produced with bulk densities ranging from just below this figure down to 12 g/litre (12 oz./ft³) or even lower. However, at bulk densities of 16 g/litre (1 lb/ft³) and below, the cell walls are only 1–2 µm thick and have almost reached the elastic limit of the polystyrene. If articles of very low density are produced by moulding the requisite weight of expandable beads in a mould of the required volume, poor fusion and excessive shrinkage normally occur. This is due to the beads expanding almost to their limit with the available blowing agent and filling the mould but without retaining sufficient expansion power to fill the voids between the expanded beads and thereby form a completely homogeneous fused mass.

After pre-expansion the beads are allowed to cool, and the residual blowing agent condenses inside the bead, thus causing a partial vacuum. At this stage the beads are extremely weak and if pressed between finger and thumb they collapse. During the maturing period air permeates through the multi-cellular structure of each bead until equilibrium with the atmosphere is achieved. The beads are then resilient and do not collapse when squeezed between finger and thumb. In this condition the beads are ready for moulding since expansion of the air within the bead during the moulding process can contribute to the pressure generated by the residual blowing agent and so ensure good fusion.

For a given density there is a minimum and maximum maturing time. The minimum time is determined by the rate of air diffusion into the beads while the maximum time depends on the rate at which the residual blowing agent permeates out to the atmosphere. If excessive loss of blowing agent occurs, the expansion characteristics during moulding are impaired. Maturing times normally range from 9 to 24 h.

When maturing is complete, the pre-expanded beads are transferred to the mould, which is then closed and heated by injecting steam. The residual blowing agent and the air which entered the beads during maturing both expand and the polystyrene is again softened. Since the beads are confined in a closed mould, the expansion in volume of the blowing agent and air within the beads causes them to distort and fill the voids between the beads. Individual beads merge into the mass, giving a coherent, non-communicating, microcellular structure. The mould is then cooled, opened and the moulding removed. An examination of an expanded polystyrene moulding will clearly show the microcellular structure described.

It will be seen from the foregoing that it is important to achieve the correct bulk density during the pre-expansion process because the only change that occurs during the moulding process is a slight expansion of the beads and their fusion in the confined space of the mould. The weight and volume of the finished moulding is hence the same as that of the pre-expanded beads used to fill the mould.

The batch-wise pre-expansion of expandable beads has been carried out in equipment as simple as an oil drum, fitted with a wire gauze support for the beads. Steam was introduced into the base of the drum and passed up through the beads. Any condensation drained through holes in the drum base. A more suitable piece of equipment for batch pre-expansion is shown in Fig. 14.1. Basically it is an autoclave, equipped with a stirrer. The beads are introduced via a pipe at the top of the autoclave while

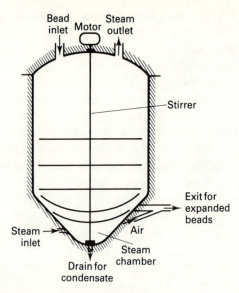

**Figure 14.1** Autoclave batch pre-expander.

steam is led in through the false bottom. The pre-expanded beads are collected from the base of the autoclave via a pneumatic conveyor. Continuous methods of pre-expansion are also available. One method is based on a free-space steam heater which consists of a stirred vertical cylinder with horizontal take-off points at various heights up the cylinder. The expandable beads are fed into the base of the cylinder by a screw feed or by a steam injector. As the beads expand they are swept upwards and out of the pre-expander. The required density is usually obtained by controlling the feed rate of the expandable beads, allowing air to bleed into the steam supply to reduce the temperature, or by varying the height at which the pre-expanded beads are removed.

Whichever method is used, whether batch or continuous, the resultant beads have to be matured (as explained earlier). Subsequent moulding of shaped articles such as contour packs or carton inserts is carried out by one or other of the following basic methods. In the autoclave method, the mould is filled with pre-expanded beads, closed, and placed in the autoclave. Fusion of the beads is achieved by steaming for a given period at 70–140 kN/m$^2$ (10–20 lb/in.$^2$) steam pressure. A convenient method of mould filling is to use an air injector. Using an air line pressure of around 700 kN/m$^2$ (100 lb/in.$^2$), the mould is completely filled with the beads. In addition, a certain amount of pre-compression occurs and this ensures that in the subsequent moulding operation, full use is made of the expansion power of the beads to fuse completely into a homogeneous mass.

Automatic moulding machines are also available, working on similar lines to a conventional injection moulding machine. The pre-expanded beads are injected into the mould by means of an air injector and then fused by steam. This pre-heating avoids condensation forming during moulding. The beads are then injected and when the mould is filled, steam at 105–140 kN/m$^2$ (15–20 lb/in.$^2$) pressure is introduced into the male section and at 210–280 kN/m$^2$ (30–40 lb/in.$^2$) pressure into the female section. The steam displaces the air from the voids between the beads and the mould is then completely closed. Steaming is then continued until fusion of the beads is complete. At the end of the steaming cycle, the steam is quickly vented to atmosphere and cooling is carried out by circulating cold water through both parts of the mould. The mould is then opened and the article ejected automatically.

Although steam heating is still the most widely used heating medium for expandable polystyrene moulding machines, other processes have been developed. In the USA, one process employs dielectric heating instead of superheated steam. One of the main advantages is that it allows the use of reinforced polyester moulds because of the lower operating temperatures. Such moulds are much cheaper than equivalent metal moulds. Another advantage is that faster production rates are possible due to reduced heating and cooling cycles.

## 14.2 Sheet extrusion

Expanded polystyrene sheet is important because it can be thermoformed to produce a wide range of trays, box inserts and boxes. Initial work on the extrusion of expanded polystyrene sheet showed that the normal blowing agents used tended to form solutions in the polystyrene at extrusion temperatures. This led to the formation of large bubbles and hence a coarse cell size. This is undesirable so a nucleating agent, such as a mixture of citric acid and sodium bicarbonate, is used. There are two main methods for the production of expanded polystyrene. The first starts with expandable polystyrene beads, i.e. beads that have been impregnated under pressure with a liquefiable gas, such as pentane. The manufacture of these beads was described in section 14.1.

The sheet is blow extruded, using a twin-screw extruder, the nucleating agents having been mixed with the beads just prior to extrusion. To avoid premature decomposition of the nucleating agents, it is essential that the expandable beads be pre-dried just before the mixing. One suitable method is to blow dried air through the beads for 2–3 h. This method of sheet production is particularly suited to low tonnage production of the order of 300–350 tonnes per annum.

The other method is direct gassing in the extruder. Ordinary crystal polystyrene is premixed with the nucleating agents and loaded into the feed hopper of the extruder. The granules are melted the first stage of the extruder barrel, the melt temperature being then about 240 °C. A liquid blowing agent is then injected into the melt, using a high-pressure diaphragm metering pump. A mixing section of the extruder screw ensures good dispersion of the blowing agent. The temperature of the melt is then reduced to about 130 °C and the melt is forced through the annular die of the blowing head. This method is the most economical for large tonnage sheet manufacture. The original liquid blowing agent was one of the fluorinated hydrocarbons, a big advantage being the elimination of fire risk during manufacture and storage. However, because of the implication of fluorinated hydrocarbons in erosion of the ozone layer, the use of pentane is now preferred.

Whichever method of extrusion is used it will be appreciated that the material leaving the extruder expands considerably with a consequent drop in density from 1040 g/litre (65 lb/ft$^3$) to about 80 g/litre (5 lb/ft$^3$). Thus, a thirteen-fold increase in volume occurs. The sideways component of this expansion would cause corrugation to occur if a slit die were used. Using an annular die the corrugations can be eliminated by blowing the tubular extrudate into a bubble, as with lay-flat polyethylene film. After the bubble has been collapsed during the haul-off operation, the lay-flat tube is trimmed on each side to produce two flat sheets that can be reeled up together or separately. This is necessary because the folded edges of the tube are a source of weakness due to the rigidity of polystyrene. Control of sheet orientation is achieved in the usual way by adjusting the blow-up ratio and the haul-off speed. Orientation should be as balanced as possible to avoid splitting of the sheet during any subsequent thermo-forming operations. The bubble is normally blown horizontally for ease of handling during start-up because the material is hard and rigid, unlike polyethylene film.

Although the density of the sheet as it leaves the extruder is about 80 g/litre (5 lb/ft$^3$), subsequent operations such as blowing-up, and stretch-ing during haul-off, tend to compress the sheet and so increase the density. In addition, the collapsing of the bubble and subsequent passage of the sheet through nip rolls also increase the density. Normally, therefore, the density of the finished sheet will be of the order of 128–160 g/litre (8–10 lb/ft$^3$). The density can be reduced by passing the sheet under an infra-red heater, at the same time supporting the sheet across its width by a roller. If the expanded sheet is reeled and allowed to stand for a few hours prior to the thermoforming process, the cells absorb air and a further reduction in density is obtained when exposed to the thermoforming heaters. This thickening effect is useful in countering the tendency for a thermoformed article to become thin at the corners.

## 14.3 Possible defects

Too low a density could result in mouldings that are too weak for the intended end use. As seen earlier, density is primarily determined at the pre-expansion stage.

Inadequate filling of the mould, too long a time between maturing of the beads and their use, and faults in timing or temperature of the moulding process can all lead to weak or 'crumbly' mouldings.

Imperfect thermoformings will often be traceable to the original sheet and some of the reasons for weakness were mentioned earlier.

# CHAPTER 15

# Printing and decoration

There are many possible reasons for printing or decorating a container and the particular reasons must be taken into account when choosing the type of ornamentation. At its simplest, it may be required only for identification but in these days of sophisticated marketing techniques it usually has to act as a selling aid as well. This does not stop at simple decoration and the package often has to advertise the latest give-away or premium offer. In addition, foodstuff containers have to carry instructions for serving or for recipes based on the contents. Finally, it may have to carry information about the contents in order to comply with legal requirements (as with pharmaceuticals and foodstuffs).

## 15.1 Background ornamentation

Before looking in more detail at printing and decoration, it is worth noting that package decoration is dependent on two main elements, namely, shape and surface appearance. Container shape is particularly applicable to plastics because of the versatility of the various conversion processes. Brand identification can very well be achieved by a particular bottle shape and with plastics there are few limitations. There are, of course, a few design factors such as avoiding abrupt changes in section and rounding off corners since these are likely sources of weakness if the container is stressed at any time.

Another method of ornamentation which does not involve any post-moulding process is self-colouring and this, again, is very easily achieved with plastics. A wide range of colours is available with most plastics and this method of ornamentation has many attractions.

It is also possible to obtain 'printing' without a post-moulding process. This is achieved by cutting into the mould surface so that the required message appears as embossing on the finished container. There are limitations, here, on the minimum size of printing obtainable and the method is not suitable for lengthy texts, such as explanations or directions for use. For brand names and simple descriptions of the contents, however, embossing can be most effective. Another method of producing an embossed surface on the container is by the use of inserts in the mould. This enables the pattern or printing to be changed without the necessity of producing a completely new mould. It is particularly useful where changes have to be made after short production runs, e.g. a change of grade. In addition to embossing, the technique of texturing the mould surface can be adapted to produce graining and other textures on the outside of the container.

## 15.2 Labelling

Labelling is another technique that is applicable to plastics containers. If the label is of the complete wrap-around type, then adhesion is between two surfaces of the label and no difficulties arise. Conventional equipment is suitable except that it may be necessary to modify the conveying system because of the light weight of the plastics bottles. Suitable equipment is, however, available from labelling machine manufacturers.

If a wrap-around label is not feasible and adhesion has to be obtained directly between label and container surface then account must be taken of the type of plastics surfaces. Very inert plastics, such as polyethylene and polypropylene may have to have their surfaces pretreated before the label is applied. Such pretreatment is also necessary before printing and the relevant methods will be dealt with later. In any case, it is essential that the label supplier is made aware of the type of surface to which it is proposed to stick the labels, so that he can supply the correct adhesive. As far as the type of design and printing is concerned, labels offer the widest range of ornamentation at modest cost but there is, of course, always the danger that they will become dislodged during storage of the filled containers and thus identification is lost.

One method of labelling which overcomes this problem is the plastics film label in the form of a band placed around the container and then shrunk-on by heat. If the band is formed from flat film, as distinct from lay-flat tubing, it can be printed so that the printing ink is between the film and the container and a scuffproof label is thus formed. Because the plastics film is printed as a flat surface, it is possible to print quite elaborate designs in one operation (as with paper labels). The films normally used for shrink-on labelling are oriented PVC, polyethylene and polypropylene.

### 15.2.1 Heat transfer labelling

Another, special, type of labelling is heat transfer labelling. Basically, this consists of the automatic and instantaneous heat transfer of a gravure

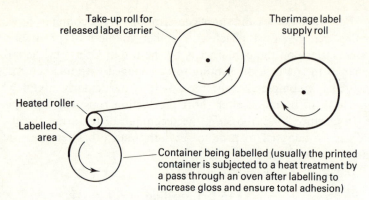

**Figure 15.1** Heat transfer labelling process.

print, from a special release-coated carrier web, to the container surface (see Fig. 15.1).

The carrier web is then rewound. The print cools rapidly on transfer and has good adhesion to the moulding. However, it is usual to pass the decorated containers through an oven directly after transfer, in order to improve the surface gloss and ensure maximum adhesion. Speeds of 30–100 articles per minute are possible.

The advantages include greater freedom of design and the fact that a multi-coloured print is achieved in one pass. In addition, the process is completely automatic and can be run by one unskilled operator. The reject rate is also lower than with silk screen or offset processes.

The main disadvantage is the high cost of equipment and of the gravure-printed labels (due to high origination costs). If very large quantities are to be printed, then the heat transfer process becomes economically viable in comparison with offset and screen printing. No hard and fast figures can be given for the minimum economic run but in general somewhere around the million mark seems to be reasonable.

## 15.3 Printing

Printing is the most popular means of decorating plastics packages and several methods can be used. Some of these are applicable to moulded or thermoformed items, blow moulded bottles and films but others are more limited. For instance, silk screen printing can be applied to most plastics items but gravure printing is usually limited to film (although it is possible to gravure print special heat transfer labels and apply them to bottles, etc., in such a way that the decoration becomes part of the pack – as with heat transfer labelling).

### 15.3.1 Pretreatment

As was mentioned, briefly, when discussing labelling, it is sometimes necessary to pretreat certain plastics prior to printing, in order to obtain satisfactory adhesion. Of the plastics commonly used in packaging, polystyrene and PVC do not require surface pretreatment but polypropylene and the polyethylenes usually do. This is because of their inert, non-polar structure which inhibits any chemical or mechanical bonding between the printing ink and the plastics surface. Pretreatment processes are aimed, therefore, at producing a surface to which the printing ink can key and they are usually based on oxidation processes.

*Flame treatment*

The flame treatment process is widely used for the pretreatment of blow moulded and injection moulded articles. It consists of exposing the plastics surface to a suitable oxidising flame for a short period of time, usually in the range 0.2–3.0 s. The flame treatment changes the plastics surface in such a way that it becomes wettable and allows a strong adhesive bond to be formed between the surface and the ink. A variety of flame pretreatment equipment is available, including linear conveyors where the article is taken past burners placed on one or both sides of the belt. In some cases the article may be rotated as it passes the flames. Where cylindrical bottles are to be pretreated it is possible to use a ring burner and drop the bottles through the flame and then through a water-cooled chute.

Both under- and over-treatment give poor print adhesion. One of the critical factors is the gas/air ratio and once the optimum has been determined for a particular operation, it is advisable to keep it constant by carefully monitoring the gas and air, by attaching meters to both supply lines. Antistatic additives, which may have been incorporated in the plastic, will have an effect that may depend on the type and concentration of additive used and on the type of flame pretreater. Some antistatic additives are only effective after flaming but in cases where the diffusivity and compatibility of the additives make them effective immediately after processing, flaming conditions are more critical. Such antistatic additives can cause trouble if used at higher than recommended concentrations.

*Electrical treatment*

The electrical treatment method is particularly useful for films where the thin gauges involved make it difficult to use flame processes. The pretreatment is achieved by passing the film between two electrodes, one of which is a metal blade connected to a high-voltage (10–40 kV), high-frequency (1–4 kHz) generator. The other electrode is an earthed roller and is separated from the high voltage electrode by a narrow gap of about 1.5–3 mm. The earthed roller is usually made from steel covered with a dielectric such as polyester film. The metal blade electrode should be slightly narrower in width (about 5–10 mm) than the film to be treated in order to prevent direct discharges to the roller. The electrical discharge is accompanied by the formation of ozone which oxidises the surface of the film, rendering it polar and receptive to inks. A certain mechanical roughening of the film may also occur due to the formation of micro-pits and this also helps to key the ink.

### Chemical treatments

Chemical treatments are based on the use of strong oxidising agents, such as chromic acid, which attack the polymer surface to form carbon—oxygen bonds. They are obviously messy and dangerous and are little used except for small numbers of awkwardly shaped mouldings which might be difficult to treat evenly by flame or electrical methods.

### Effectiveness of pretreatment

A simple test for detecting whether or not the surface has been pretreated is to dip the object in water. An untreated surface repels the water immediately, whereas a treated one retains a film of water for up to several minutes.

One widely used test to determine the efficiency of pretreatment uses the strength of film adhesion as a yardstick. A length of pressure-sensitive tape is firmly pressed on to the surface and the peel strength is then measured with a tensometer.

In general, the time between the pretreatment and the printing process should be kept to a minimum. This is because the effect of the treatment diminishes with time and because the surface, after pretreatment, is sensitive to handling and dust pick-up.

### 15.3.2 Printing methods

Some of the printing processes suitable for printing on plastics are described in the following pages.

*Silk screen*

This is essentially a stencil process and was originally based on a silk screen. Today, the screen is more likely to be made from fine, stainless steel wire or nylon or polyester mesh. The screens usually have about 8–12 apertures per millimetre (200–300 per inch) and they are prepared by photographic methods, being porous only in the areas where decoration is required. The screen is supported on a frame which holds the ink supply and keeps the screen taut. A rubber squeegee is drawn across the screen and the ink is forced through the porous areas of the screen on to the plastics surface (Fig. 15.2).

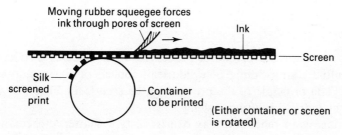

**Figure 15.2**   Silk screen printing process. (Source: Briston and Katan: *Plastics in Contact with Food*, Food Trade Press, London.)

The screen is moved out of contact with the moulding and the printed area is normally stoved to dry the ink film. The 'lay down' of ink is heavy, the print is even and sharply defined, and either a matt or gloss finish can be obtained. The process is widely used for the decoration of blown bottles, injection mouldings and vacuum formed articles at speeds in the range 100–4000 items per hour. Fast-drying inks which allow drying of the ink film to be carried out by cold or hot air or by rapid flaming, have been developed during the past few years.

Advantages of silk screen printing include low cost of equipment and screens, short changeover time and the easy training of operators. It is also an economically attractive process for short runs, while also being adaptable to longer ones, using more sophisticated equipment.

The main disadvantage of silk screen printing is the fact that, for multi-colour printing, it is usually necessary to use a separate screen for laying down each colour, with suitable drying equipment between each station. If the colours in the final design do not overlap, however, it is possible to use a split screen to give a multi-colour effect in one pass.

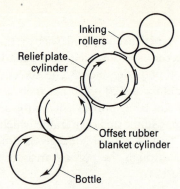

**Figure 15.3** Principle of the dry offset process. (Source: Briston and Katan: *Plastics in Contact with Food*, Food Trade Press, London.)

*Dry offset*

Another name for the dry offset process is offset letterpress. It is used when multi-colour printing of cylindrical, conical or flat mouldings is required. The principle of the process can be seen from Fig. 15.3 which shows a cylindrical printing system.

The image is in relief and may consist of type matter, etched zinc or copper blocks or various rubber or plastics alternatives. A collection of such items for printing from the flat is known as a forme. A cylindrical replica of the whole forme is known as a stereo. Stereos are not easily made and the modern trend is to prepare a thin flat plate and wrap it around a plain cylinder.

Tone reproduction is achieved by breaking the picture into a large number of small dots. These are equally spaced but the size varies with the tone. Light tones are achieved by using very small dots whereas in dark tones the dots are larger and merge into one another. The rubber blanket cylinder is an important part of the dry offset process because it picks up the image from the raised plate and transfers it to the article being decorated. Accurate reproduction of the image depends on a faithful reproduction of each dot and this depends on accurate control of the pressure between the relief plate and the rubber blanket. Excess pressure between plate and blanket leads to an increase in dot size because the rubber then picks up ink from the edges of the raised image.

Where multi-colour printing is required, the rubber blanket cylinder is made to contact a number of relief plate cylinders each of which carries part of the design and is in contact with the inking roller of the appropriate colour. As with silk screen printing, the mouldings have to be stoved before filling or packing. Printing speeds vary in the range 500–6000 items per hour.

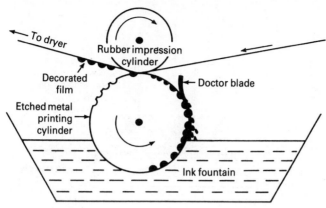

**Figure 15.4**  Gravure printing process. (Reproduced from Briston, *Plastics Films,* 3rd edition, Longman, London.)

The main advantage of the dry offset process is its low cost in comparison with other multi-coloured processes. Its disadvantage is that the ink film is thin in comparison with that obtained with silk screen printing and hence the gloss of the print is largely dependent on the surface finish of the moulding.

### Photogravure

The process of photogravure is used for film and sheet, but as it is also used for printing the heat transfer labels, mentioned earlier, it is briefly described here. The process consists of revolving an etched metal roller in an ink reservoir. The principle is shown in Fig. 15.4.

The ink is held in the recesses of the etched design, excess ink being removed by a doctor blade. The film is pressed against the etched roll by a rubber impression roller and the image thus transferred. With multi-colour printing, drying between each colour is essential.

The main disadvantages of photogravure printing are the high initial costs for the etched metal rolls and the printing speeds, which are somewhat slower than those obtainable with flexographic processes. Normal photogravure speeds are of the order of 20–120 m/min (65–390 ft/min). The main advantage of photogravure printing is that it can produce such high quality, multi-colour, fine detail printing.

### Hot foil stamping

Hot foil stamping is an expensive method of ornamentation used for decorating certain cosmetic containers such as polystyrene talc

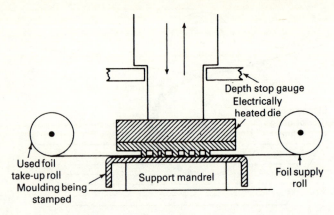

**Figure 15.5** Hot stamping process. (Reproduced from Briston, *Plastics Films*, 3rd edition, Longman, London.)

bottles and powder compacts where high quality and definition are needed.

It consists of the fusion and release of a special coating from a heat-resistant carrier tape and its transfer on to the moulding to be decorated (see Fig. 15.5). There are two effects possible with this process, namely, an inlaid or embossed effect and the top stamped effect. The stamping foil consists of a carrier tape (polyethylene terephthalate, cellulose film or glassine paper), a release layer, which is a wax, melting at a predetermined temperature when brought into contact with a heated die, a metal or pigmented film coating and an adhesive layer, which acts as a bond between the imprint and the moulded plastic. There are three process variables, namely, die temperature, contact pressure and dwell time. The actual conditions for a particular plastic cannot be given since all three variables are interrelated and good blocking can be obtained using more than one set of conditions. These also depend on the machine and the foil used.

Machines range from hand-operated units to fully automatic ones, with outputs from 500 to 5000 prints per hour. In spite of the high cost there are many advantages; for instance, the print is dry and the article can be handled immediately after printing. Also no pretreatment is required, even with polyethylene and polypropylene. The print has good opacity, the finish can range from gloss to matt and a wide colour range is available. In this respect, it should be noted that some decorative effects, such as metallic finishes – gold and aluminium, for instance – cannot be achieved by other techniques. Fine quality printing with good accuracy is also obtainable.

*Electrostatic printing*

The electrostatic technique has one important property, namely, the fact that it enables printing to be carried out on awkwardly shaped articles not easily printed by conventional means. It utilises an electrically charged stencil consisting of a fine-mesh, electrically conducting metal screen on which the non-image areas are masked. The other component of the system is a conductive backing plate of the same relative size and shape as the screen and placed roughly parallel to it. A finely divided powdered ink is applied to the outside of the screen where it takes on the charge of the screen and is attracted, though the openings, towards the oppositely charged back plate. The ink sticks to the plate in a pattern that faithfully reproduces the design in the printing screen. In the same way, any article that is interposed between the plate and screen will also be printed. The image can be fixed by heat, solvent or vapour. The intercepting medium which is to be printed can consist of any material provided it does not interfere with the electrical field. In this respect, articles of high capacitance, such as those made from polystyrene or polyethylene, can give trouble unless the static charge on the surface is first dissipated. The general principles of the process are shown in Fig. 15.6.

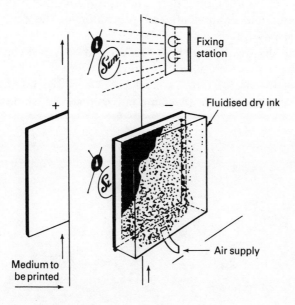

**Figure 15.6** Electrostatic printing. (Reproduced from Briston, *Plastics Films*, 3rd edition, Longman, London.)

*Tampo printing*

Tampo printing is an offset gravure process that uses a deformable ball or cone, made of rubber or gelatine, which transfers ink from a flat gravure plate to a container. This process has the advantage that it can print in more than one plane at once so that it is another method for decorating irregular or multi-contoured surfaces. Because it also uses half tones, good multi-colour effects can be achieved.

*In-mould decoration*

In-mould decoration is a process that has been widely used for the decorating of thermosets and is now being applied to the decoration of injection moulded, blow moulded and thermoformed plastics.

A reverse printed film (some 75–125 µm thick) is placed in the mould, before the moulding process starts. The decoration thus becomes permanently embedded in the moulding. If a transparent film is used it must be of roughly the same refractive index as the moulding. The printing process used can be gravure, flexographic or silk screen, according to the effect desired. The label can be held in place in the mould by gravity, mechanical clamping, vacuum or by inducing a static electrical charge. One advantage of the process is that freedom of design is increased because of the number of different printing effects which can be used. Other advantages include elimination of postmoulding decoration processes, with savings in floor space, labour and equipment, and protection of the decoration from chemical or physical attack.

The main disadvantage is high cost, due to increased cycle times and the fact that it is more difficult to reprocess reject mouldings due to the presence of the decoration. Today there is renewed interest when the container and label are of the same polymer, so the whole item can be recycled.

CHAPTER 16

# Choice of forming method

The choice of forming method may be dictated by economic or by technical factors. Looking first at some of the technical factors we find that, in general, injection moulding cannot be used to mould objects having re-entry curves or reverse tapers. This is because it would be difficult, if not impossible, to remove the moulded object from the mould. It is, however, possible to mould slight undercuts in certain materials, though not in others. Undercuts have been moulded in polypropylene screw caps, for example, into which may be inserted decorative items such as imitation jewels (made from polystyrene). Polypropylene is a sufficiently resilient material for this purpose but a similar attempt using polystyrene for the cap material would lead to fracture when the cap was removed from the mould (see Fig. 16.1). The screw threads inside a closure are also effectively undercuts, but the item is removed from the male component of the mould by unscrewing it.

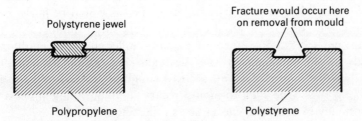

**Figure 16.1** Undercuts in polypropylene and polystyrene.

Another restriction of the injection moulding of tubs or cups is the wall thickness involved. There is a lower limit to the wall thickness achievable by injection moulding, dictated by the flow properties of the particular thermoplastics being moulded. For articles where wall thicknesses below this limit are required, thermoforming would be the appropriate process.

Similar restrictions on reverse tapers and re-entry curves also apply to articles thermoformed from sheet but there is a little less restriction here, because the combination of a fairly thin sheet and a resilient material would allow a greater degree of freedom. Other limitations may be imposed by the material to be thermoformed. It is difficult to thermoform nylon sheet, for example, because nylons usually have fairly sharp melting points. This makes temperature control a critical matter.

Hollow components with apertures smaller in diameter than the body are usually made by blow moulding (either extrusion or injection blow moulding being suitable).

For very large items, such as tanks and intermediate bulk containers, the rotational moulding process is usually the most suitable. As was seen earlier, mouldings are produced without the use of high pressures and hence without moulded-in stresses.

There are, however, many occasions when two or more forming methods are technically feasible and factors such as capital expenditure, production costs, mould costs and the length of run then become of paramount importance.

Taking injection moulding as an example, the main factor is the high cost of moulds. This is because of the high pressures involved and the fact that the inside surfaces are usually chromium plated. Capital costs are also high. The length of run has to be appreciable, therefore, in order to amortise the costs involved.

Mould costs are not so high for blow moulding as the pressures involved are much lower (i.e. those of the compressed air used) and the moulds do not need to be so massive.

Thermoforming costs vary considerably according to the size of the moulding and the numbers off required. For reasonably small runs of, say, a blister pack, mould costs and capital costs would be fairly low. If, however, we consider a requirement for very large numbers off (of a yoghurt pot, for example) then the picture is different. Large, multi-cavity moulds would be necessary and the high-speed, automatic equipment would also be high in cost. Such costs would be justified, of course, for very long runs. The melt-to-mould rotary thermoforming process also saves money by reducing energy costs (see section 12.5.1).

Before leaving the subject of the economics of forming it is worth looking at the influence of the moulding material. The intrinsic price per kilogram of a polymer is often not the best guide to its economies in use. What has to be examined is the cost per unit in terms of the properties required in the finished moulding. In the case of a blow moulded bottle or a thermoformed tray it may be the barrier properties towards water vapour or gases that are important and it is the cost per barrier unit that matters.

Processability is another factor that can reverse the apparent economics of two different materials. This is more likely to apply to different grades of the same material as the material choice itself will probably have been made on the grounds of the end-use properties required. In injection moulding, for example, a high flow grade may prove cheaper overall, provided that the other properties are satisfactory. For thermoforming the grade with the greatest rigidity at elevated temperatures might be more economic because the time taken for the heated forming to cool sufficiently for it to support its own weight is the governing factor. The decreased cycle time in each case can lead to substantial savings in a long run.

One final point. Polymer granules or powder are bought by weight but the mouldings are made by volume so that a 10% difference in density means a 10% difference in the number of mouldings per kilogram.

*PART 3*

APPLICATIONS

# CHAPTER 17

# Food and drink

There are few food products that are not packaged in some way, so that food packaging accounts for a large proportion of all packaging. In the USA, it has been estimated that approximately half of all packaged goods are foodstuffs of one sort or another.

By their nature, foodstuffs are biological and can deteriorate in a number of ways. Thus, they may lose their nutritive value, change in flavour, colour or texture and may even become a toxicological hazard. Deterioration of food can occur in four main ways, namely, biochemical, enzymatic, microbiological and physical. Packaging has to protect against these deterioration mechanisms by maintaining certain conditions inside the package and by preventing ingress of undesirable elements of the environment.

As far as plastics are concerned a distinction needs to be made between protection against microbiological changes and chemical or physical effects. Practically any good quality plastics container is impermeable to micro-organisms so that if a sterile foodstuff is packaged in a properly sealed plastics package, no further microbiological changes are likely to take place. On the other hand, chemical or physical changes can continue to occur either because of reactions within the foodstuff or because the plastic is not impermeable to gases or water vapour. The latter is particularly important and much of the function of food packaging is to prevent loss or gain of water from the product.

A list of foodstuffs already packaged in rigid or semi-rigid plastics containers of one type or another would be extremely lengthy and not particularly informative. The various areas will be examined, therefore, with a view to illustrating the factors of choice involved.

## 17.1 Milk

Milk is probably one of the largest potential markets for plastics bottles and in some countries plastics already have a large share. In the UK the two main contenders are low- and high-density polyethylene. Thin-wall LDPE bottles have been produced by the form/fill/seal bottle blowing technique described in section 11.4.2. They were not cheap enough to be used as regular one-trip containers and were used, mainly, for milk sold over the counter. This market has now been taken over by HDPE and both 2 and 4 pint containers of milk are a common sight in supermarkets. HDPE bottles have the advantages over glass of light weight and a smaller overall volume (to the extent that a $1\frac{1}{2}$ pint HDPE, square-section bottle takes up about the same space as the 1 pint glass bottle.

The doorstep delivery, which is still used to a very great extent in the UK (although the percentage is falling: it is now about 60% of milk sales), favours the use of returnable (glass) bottles. However, in the USA, doorstep delivery is very small and the market was dominated, first by wax-coated board cartons and later, by LDPE-coated cartons. HDPE bottles are now widely used as non-returnable containers, usually in large sizes such as $\frac{1}{2}$ and 1 gallon capacity. The light weight and resistance to breakage of HDPE are attractive properties when the milk has to be carried instead of being delivered. Polycarbonate, too, has been used in the USA as a returnable container for milk, with a claimed trippage of around 100.

Milk packaging, in particular, seems bedevilled by various groups of people with special pleadings and the choice is not at all straightforward. Proponents of returnable containers claim that one-trip containers contribute greatly to the high volume of household waste and use a disproportionate share of the Earth's resources. Opponents, however, claim that returnable containers can be non-hygienic. Work on deliberate contamination of glass, polyethylene and polycarbonate containers, reportedly showed that in repeated use, polyethylene containers retained volatile substances which affected milk taste, some polycarbonate containers were damaged by chemicals and those that were undamaged showed an affinity for organochlorine compounds. It was also found that even after unobjectionable repeated use, milk fat absorbed by plastics containers (particularly polyethylene ones) becomes rancid. Polycarbonate is not as bad as polyethylene in the latter respect although there are still the problems of possible contamination through alternate use by the consumer. In Germany, where polycarbonate bottles have also been used, the bottles are inspected visually and then by an automatic measuring device that takes an air sample from every bottle. The air is analysed by gas sniffer techniques to detect aromatic hydrocarbons and any contaminated container is automatically rejected.

It is interesting to note that milk was the first foodstuff to be packaged in the now ubiquitous bag-in-box. First introduced in the 1960s it was used mainly as a bulk pack for milk. This milk was sold through the Milk

Marketing Board's chilled cabinet dispensers in outlets such as milk bars, restaurants and cafeterias. Milk is still being packaged in 3-gallon bag-in-box packs and is now dispensed in public houses in England and Wales!

Milk intended for long-life storage is subjected to high temperature for a short time and can then be aseptically packaged in containers made from a laminate of LDPE/paperboard/aluminium foil/LDPE. In aseptic packaging the product must be aseptically processed, remain commercially sterile and be filled into a pre-sterilised container in an atmosphere free from unwanted micro-organisms. The container sterilisation in this instance is achieved by the action of hydrogen peroxide and heat on the inner polyethylene surface of the laminate.

## 17.2 Other dairy products

One dairy product that owes its tremendous growth to plastics is yoghurt. For many years yoghurt was a fairly low volume product, catering mainly to what was then a small health food market and was packaged in glass containers similar to small milk bottles. When fruit-flavoured yoghurts were introduced, the resultant marketing strategy demanded a container light in weight and having a bright and attractive appearance that would help sell the new product. It also had to be capable of being filled at high speeds. High-impact polystyrene tubs satisfied these requirements in all respects. The fruit could be portrayed in mouth-watering colours while the aluminium foil closure also added to the sales appeal. The wide mouth of the tubs enabled them to be filled at high speed and the finished pack was easily displayed on supermarket shelves.

Similar tubs are used for the packaging of dairy cream and a huge market now exists for high-impact polystyrene in these applications. Such tubs can be made by either of two processes, namely, thermoforming and injection moulding. Originally, the major process was injection moulding, for a number of reasons. The injection moulding process was the cheaper process although thermoformed tubs could be made with thinner walls and hence save on material costs. This was because of a lower limit on wall thickness with injection moulding due to flow limitations within the mould cavity. However, early yoghurt formulations required the use of a fairly rigid container because repeated flexing caused the product to separate and coagulate. In addition, the early thermoformed containers were not easily de-nestable and this caused hold-ups on the filling line. Finally, the rims of themoformed tubs were not rigid enough to withstand the pressures exerted by the closing equipment then in use.

Thermoforming has always been the method used for the manufacture of vending drinking cups, however. They have to be cheap, de-nesting does not have to be carried out at high speed and their thin walls do not have to stand up to handling on high-speed filling lines.

Developments in thermoforming (and changes in the formulation of yoghurts) have since changed the situation. Adequately rigid thermoformed tubs are now available, while the development of high flow grades of polystyrene have led to reductions in the minimum wall thickness attainable by injection moulding. Both types of tub are now used in similar applications.

Yoghurts are also packaged in pots produced on thermoform/fill/seal machines instead of in pre-made pots. Such pots can be as much as 25% cheaper to produce. High barrier coextrusions are used and these give longer shelf-lives of up to 25%.

Ice cream is another dairy product that is widely packed in plastics containers, particularly the specialist ice creams and the larger (1, 2 and 4 litre) sizes of ordinary ice cream. These large containers are usually made by injection moulding, using HDPE, polypropylene or high-impact polystyrene. The lids are usually made from LDPE or high-impact polystyrene and are of the snap-on variety. A stick-on paper label on the lid carries all the necessary printing and decoration. The bodies are sometimes self-coloured.

## 17.3 Beverages

The use of plastics containers for beverages should be considered under two main headings, namely, pressurised and non-pressurised. Pressurised beverages contain large volumes of carbon dioxide, dissolved under pressure, and so make much greater demands on the strength and barrier properties of the container.

### 17.3.1 Non-carbonated beverages

Non-carbonated beverages include the so-called fruit squashes and crushes,- fruit cordials and the non-sparkling wines. In the UK, the most common plastics material for fruit squash containers is PVC. This material is also used in France for many of the cheaper wines. It is clear, is a good barrier to oxygen and is resistant to essential oils. Coextruded bottles have also been used in the fruit squash market where particularly good retention of essential oils is required. A typical bottle is one based on the three-ply laminate of HDPE and ethylene/vinyl alcohol copolymer (EVOH), the latter being the centre layer. The HDPE supplies impact resistance and rigidity at a lower cost than PVC while the EVOH provides the barrier to essential oils and oxygen. Laminates having as little as 7% EVOH have been used and reputedly have a resistance to oxygen ten times that of PVC. On balance, PVC remains the material of choice, mainly due to price. Although only a thin layer of the more expensive EVOH is necessary to give equivalent barrier properties, the straightforward extrusion blow moulding of PVC is a cheaper process than coextrusion.

A novel container that has been used in France as a replacement for the returnable wine bottle used in the French institutional market is the Tetra King (now discontinued). This container was formed from two webs of a laminate comprising an inner layer of polystyrene, a 1 mm thick layer of expanded polystyrene and an outer layer of polystyrene. It consisted of two parts: a half-round body section and a section that forms the top, bottom and back of the package. Available sizes ranged from 200 ml to 1 litre.

In one version a hole is punched in one end and is then covered with a peel-off strip. A straw is taped to the container for use when the hole is open. This version has been used in the UK for the packaging of milk shakes. In another version, one of the half-round ends was scored through in such a way that it can be folded back to form a full aperture opening. This version was aimed at the packaging of viscous products such as yoghurt. A spoon taped to the container completed the image of convenience.

One expanding market for poly(ethylene terephthalate) – PET – is that of spirits such as gin, whisky, brandy, etc., particularly in the field of airline travel and in duty-free shops. The 50 ml 'miniature' PET bottle, in particular, has had a great deal of success. The major benefit, here, is a weight reduction of about 80% compared with glass. The savings in fuel derived from a changeover to PET have been estimated as about £16 000 per year for a typical passenger airliner.

Similar PET markets are also opening up for wine although there is still some consumer resistance to wine in plastics. However, this resistance is being overcome as the growth of wine in bag-in-box shows. Two main factors account for this growth. One is the convenience appeal of a 3-litre box *vis-à-vis* four 75 ml bottles, both from the carry-home aspect and the space utilisation at home. The second factor is the collapsibility of the plastic inner bag which ensures that no air comes into contact with the wine remaining in the container during use. Laminates used for the inner bag include LDPE/metallised polyester/EVA, LDPE/ionomer/nylon and laminates based on EVOH. The bag-in-box has also been used for fruit juices, utilising both hot fill and aseptic techniques. Other beverages packed include sherry, cider and perry.

One increasingly large market for plastics bottles is that of mineral waters. The material of choice is between glass, PVC and, most recently,

PET because of its clarity and sparkle. PVC is still the most widely used in France, which is the biggest supplier of mineral waters.

### 17.3.2  Carbonated beverages

Carbonated beverages include beers, stouts and lagers, together with soft drinks such as lemonade and colas. Gas barrier properties are obviously of particular importance for the packaging of carbonated beverages, as are tensile strength and resistance to creep. Other factors that have to be considered are clarity, taint and toxicity. The suitability of a number of thermoplastics for the packaging of carbonated drinks has been the subject of at least one in-depth study. This looked at the criteria mentioned earlier, in addition to certain performance criteria. These performance requirements were:

1. Creep resistance: the increase in bottle diameter to be less than 3% after 90 days.
2. Impact resistance: bottles should withstand a 1–2 m drop on to a hard surface, a test that would eliminate most glass bottles!
3. Gas permeability: the loss of carbonation after 90 days to be not more than 15%; the ingress of oxygen after 90 days to be less than 25 ppm.

Taking these criteria as a basis, poly(acrylonitrile) has excellent barrier properties but its resistance to creep and to impact are barely adequate. PVC is deficient in gas barrier properties for smaller bottles (where the surface area to volume ratio is high) when compared with poly(acrylonitrile). This is true in spite of the fact that the lower price of PVC allows it to be used in thicker wall sections. PET is also borderline in barrier properties but it has good creep resistance and excellent impact resistance, making it suitable for use in bottles of 1 litre capacity or more. This is reflected in the fact that the most common PET bottle for carbonated soft drinks in the UK is the 2 litre.

Beer and soft drinks differ in carbon dioxide content, some typical figures being 8 g/litre for colas, 6 g/litre for beer and 3 g/litre for citrus and other fruit drinks. Carbonation pressures are more often quoted in volumes where 1 volume is approximately equal to 2 g/litre. The packaging of beer has one extra problem over the packaging of carbonated soft drinks inasmuch as the presence of oxygen can cause haziness in the beer. The problem of providing the extra barrier properties necessary for the packaging of beer was first solved by coating the PET bottle with poly(vinylidene chloride) (PVDC). Commercial shelf-lives of 4–6 months were claimed, the limiting factor being oxygen attack on the beer leading to a taste change, as well as the haziness mentioned earlier. The degree of attack is dependent on factors such as the quantity of oxygen and the composition of the beer, in particular the sugar content. One problem with a PVDC coating is mechanical damage, and this technique is no longer used. Barrier polymers are more effectively incorporated using coinjection stretch blow moulding. Barriers used include EVOH and MXD 6 amorphous nylon.

The highest resistance to internal pressure is given by a sphere. The nearest practical approach to this, in container design terms, is the familar torpedo-shaped bottle, i.e. a cylindrical bottle with rounded shoulder and a hemispherical base. A high-density polyethylene or polypropylene flat-bottomed cup was originally snapped over the rounded base of the bottle to give the necessary stability. One alternative to the hemispherical bottom plus a base cup is the so-called petalloid base. This has five or six separate curved protuberances to give stability and it obviates the need for a separate base cup. It is now widely used in the UK for carbonated soft drink bottles.

One very interesting approach to the packaging of beer was the Rigello Pack. This also utilised a cylindrical body and a hemispherical base but it was formed from Barex (an acrylonitrile copolymer). A neck and a cone-shaped shoulder were also formed from Barex and were then welded to the body. The acrylonitrile-based Barex was satisfactory as a barrier to both carbon dioxide and oxygen and its deficiencies in terms of creep and impact strength were neatly overcome by enclosing the plastics body in a spirally wound paperboard tube with an outer surface of gravure-printed aluminium foil. The sleeve was rolled inwards at the bottom to make the base smooth, stable and moisture resistant. The Rigello pack was used in Sweden and the UK for the packaging of beer for some years and was competitive with its main rival, the can. It was discontinued when can prices gradually came down coupled with a devaluation of the Swedish Kroner which made the imported Barex too expensive to compete.

## 17.4 Coffee

Early attempts to package coffee in plastics containers were based on PVC but to obtain a reasonable shelf-life it was necessary to use uneconomically thick container walls. Coating with PVDC gave better results but was still too expensive. Wide-mouth PET bottles have also been tried. Sealing at the jar neck presents no difficulty but the water vapour barrier properties are still borderline for instant coffees.

Coffee has to be protected against loss of aroma and pick-up of oxygen which causes oxidation of the coffee. The answer is likely to be found in the realm of coextruded jars, probably based on EVOH as the barrier layer.

## 17.5  Cocoa and drinking chocolate

Most cocoas and drinking chocolates are still packed in tins or tin/fibre-board composites but one plastic/paperboard drum was also used for a while. This was the Vitello which consisted of an internal plastics cup thermoformed from a polystyrene/HDPE/polystyrene laminate. The body drum sleeve was rolled inwards – at the bottom, to give a smooth, stable and moisture-resistant base, and at the top, to give a firm opening to take the HDPE snap-in lid.

This type of container is only suitable for non-hygroscopic powders like cocoa and drinking chocolate and so had a rather limited market.

## 17.6  Oils and fats

One of the earliest attempts to package vegetable oils in plastics bottles utilised HDPE. The particular oil used suffered from a tendency to separate out in cold weather, with visually undesirable results, so that HDPE's lack of clarity was an advantage. HDPE was not a good enough oxygen barrier (in 1 litre sizes) and the experiment was discontinued. It is interesting to note that many years later, 25 litre containers for edible oil are being blow moulded from HDPE.

PVC is now the main material used for retail sizes and, in fact, edible oils consituted one of the earliest large-volume markets for blown PVC bottles. Some of the earliest bottles were made by the main French producer of edible oils who wanted to develop a new pack for his product. The company even developed special blow moulding equipment and special high-impact PVC grades to do the job. PVC was originally the only polymer with the necessary clarity and oil resistance at an economic cost.

More recently, edible oils have been packaged in PET bottles. According to one manufacturer of edible oils who changed from PVC to PET, the new bottle has the advantages of strength and clarity and costs no more than the old one. Although PVC cost only 80% of the raw material price, the weight of the PET bottle was reduced from 42 g to 33 g for a 1 litre bottle. In addition to being stronger than PVC, the PET bottles are said to be less susceptible to embrittlement through age or low temperature. The injection moulded necks give better sealing than the cut PVC ones.

Turning now to other fat products, soft margarines are now sold almost entirely in plastics tubs, usually thermoformed ones. High-impact poly-styrene, ABS and PVC all have the necessary fat resistance and all have been used. In Holland, however, polypropylene took over from ABS, the tubs being manufactured by the solid-phase pressure forming technique. As was mentioned earlier (Chapter 12), polypropylene is difficult to form by normal thermoforming techniques but solid-phase pressure forming enables the superior properties of polypropylene to be used to good advantage. The advantages of polypropylene over polystyrene in the field of thin-wall tubs include higher softening point, higher impact strength and lower cost per unit volume.

Another fat product is peanut butter and this has also been packed in plastics jars. One material used is a modified acrylic. The material provides the necessary high oxygen barrier (protection against rancidity), resistance to oils and a high enough softening point for hot filling.

## 17.7 Jams and preserves

The use of plastics for jams, preserves, etc. is largely confined to single-portion packs such as those used in restaurants, railway buffet and restaurant cars and similar establishments. The earliest examples were injection moulded pots made from crystal polystyrene. This material had adequate impact strength for this use because the pots were small and had low residual stresses. Hot filling was also not a problem because with such small quantities cooling was rapid. With larger jars, hot filling causes polystyrene to distort

Portion packs of jams, marmalade and honey are now often manufactured by thermoform/fill/seal techniques using PVC sheet for the containers. The lidding material is aluminium foil or metallised PET which can be attractively printed.

One larger size pack is a blow moulded, 25 litre container for golden syrup, using HDPE.

## 17.8 Tomato ketchup

One of the first food products to utilise the benefits of coextruded bottles is tomato ketchup. A good oxygen barrier is needed for tomato ketchup because oxidation causes darkening of the product. The bottle is a five-layer one with EVOH as the barrier layer. The usual make-up is polypropylene/tie layer/EVOH/tie layer/polypropylene. The squeezability of the plastics bottle aids dispensing of the somewhat viscous product. A similar bottle is now also used for brown sauce and for some ice cream toppings. One Danish manufacturer of tomoto ketchup is packaging his product in a five-layer coextruded bottle that incorporates polycarbonate as the outer ply. It is claimed that polycarbonate's clarity and stiffness give the appearance of glass while still offering squeezability. Polycarbonate can also be autoclaved as well as withstanding low temperatures down to $-20\,°C$.

## 17.9 Meat and meat products

The bright red colour of fresh meat depends on the presence of oxygen. The natural colour of the meat pigment myoglobin is purple and it is oxymyoglobin that gives the bright red colour expected by the customer. For retail sale, red meat is packed in thermoformed polystyrene or expanded polystyrene trays overwrapped with a specially formulated PVC to give the high permeability to oxygen which is necessary to retain the bright red colour. The trays themselves are usually embossed internally in order to hold any blood present, by capillary action.

Cooked meats such as bacon need high gas and water vapour barrier properties and are usually vacuum packed in flexible laminated pouches. However, a rigid, retortable pack has been developed in Sweden for the packaging of meat products, including Frankfurters. The container, known as the Letpack, was originally rectangular in shape, with rounded corners. It has since been relaunched in cylindrical form. The body is a seamless tube of polypropylene, overwrapped with a pre-printed laminate consisting of polypropylene/aluminium foil/polypropylene. The top and bottom caps ends are injection moulded from the same material and are welded on, using high-frequency welding. The top end is fitted with an easy-to-open pull top.

The filled pack is able to withstand retorting temperatures and pressures while the coated foil wrap-around label gives an excellent printing surface. The thermal processing time is about the same as for the can. The Letpack has been used by a co-operative in Sweden and by a meat processor in West Germany, and is today (1993) used in the UK for a pet food.

## 17.10 Miscellaneous food products

Thermoformed expanded polystyrene trays have already been mentioned when discussing the packaging of fresh meat. Such trays are used for the packaging of apples, tomatoes, etc. Extruded expanded polystyrene sheet has a smooth, satiny surface that gives a clean, hygienic appearance. It also has good cushioning properties that minimise any bruising of the fruit.

Expanded polystyrene sheet was also widely used for many years for the thermoforming of egg pre-packs until environmental pressures resulted in reversion to moulded pulp. The cushioning properties are obviously important here, as is the surface appearance. It is possible to design such pre-packs so that the eggs are partially visible (without opening the pack) – an important factor for quick identification of the colour of the eggs (brown eggs are still favoured by some customers). Complete visibility of the eggs (so that damaged eggs can be seen) is given by another pack –

one thermoformed from high-impact polystyrene. Although the sheet is not intrinsically a cushioning material, the design of the pack is such that direct contact with the eggs is minimised. The packs are also stackable.

Basic polystyrene is used for the injection moulding of a range of cylindrical pots for products where visibility is useful, one example being dried herbs. The closures are snap-on ones (with a tamper-evident tear-off strip) moulded from LDPE.

Finally, the blow moulding process is used for a wide range of containers other than bottles. Several different shapes have been utilised for the packing of table salt, for example. The preferred material is HDPE because of its ease of moulding, stiffness and good water vapour barrier properties. Another example is the blow moulded barrel-shaped container used for some mustards. Low/medium-density polyethylene is used in order to supply the flexibility necessary to dispense the mustard.

LDPE is also used for 'novelty' blow moulded containers, including tomato shapes for tomato ketchup, lemon shapes for pure lemon juice and 'hot dogs' for mustard.

One interesting, multi-material container, which had a wide range of potential uses, was the Kepak or Airfix container. This was developed in the UK but was never widely used. However, a number of interesting uses have been developed in Japan. Basically it consists of injection moulding a skeleton framework in a mould that contains a printed label. Conventional injection moulding equipment is used and the label is fed automatically from a magazine, shaped round a mandrel and inserted into the mould cavity. The label blanks can be made from paper, fibreboard, plastics or aluminium foil while the moulded skeleton can be made from a wide range of plastics. A large selection of possible combinations is available, therefore, to meet most performance requirements. The label blanks, which produce the container wall, can be printed in the flat. This widens the choice of printing processes and gives better results than can be achieved when printing a finished container.

Perhaps the best-known application in the UK was a pack for a dried meal product which is reconstituted by pouring boiling water into the pack. The structure used for this application was a HDPE skeleton with blanks made from an aluminium foil/HDPE laminate. This gave the required rigidity and resistance to boiling water. In this instance the label blank was unprinted, the necessary information being printed on the aluminium foil lid.

A new product area that has been opened up by the use of plastics is that of ready meals and snacks, suitable for microwave cooking. The growth of demand for such foods is due to a number of demographic trends. These include an increase in the number of women at work, an increase in single person households, changing lifestyles within the family (a move away from fixed meal times to allow for more outside activities) and a trend to eating 'on the move'. The overall needs thus created include those for greater convenience ease of preparation and, in some instances, single portion packs. The microwave oven provides the convenient method of heating and an appropriate package must meet the following requirements:

1. be a barrier to oxygen, moisture and micro-organisms,
2. be able to withstand processing and storage conditions,
3. be fully compatible with the product in terms of flavour and chemical extraction,
4. possess suitable mechanical strength and resistance to abuse,
5. be cost effective, and
6. have a good appearance.

One successful entrant in this field is the Lamipac container (trade name of CMB Packaging) which is thermoformed from coextruded sheet. A typical container make-up (from inside to outside) is polypropylene/tie layer/PVDC/tie layer/reclaim material/polypropylene. The reclaim layer consists of reground skeletal waste generated in the thermoforming process. EVOH can be used as a replacement for PVDC in aseptic packaging applications but its oxygen barrier performance is greatly reduced when moist and it can take a significant amount of time to dry out after retorting and to recover its barrier properties.

The containers available are basically of two types. The most common is a shallow tray, closed with a heat-sealed diaphragm. A more recent development is the 'Supabowl' where an easy-opening metal end is double seamed to the thermoformed bowl.

## 17.11 Transit containers

The packages described so far are basically all retail packs although the bag-in-box concept has also been used for packs ranging up to 700 litres (for fruit and vegetable concentrates and UHT milk). Rigid and semi-rigid plastics containers are also widely used for transit packs. Plastics boxes to take 6 kg of fruit such as cherries and grapes have been in use for some years in Italy and France. These boxes are steam moulded from expandable polystyrene beads and have a density of around 40 g/litre ($2\frac{1}{2}$ lb/ft$^3$). They stand up well to the rigours of international transport and have the seal of approval of the French National Railways. Use has been made of the greater flexibility of design inherent in moulded plastics boxes. Thus, a box designed to hold eight small melons has special indentations moulded into the base of the box and the melons are held apart, to prevent bruising, without the use of the normal tissue paper. This represents a saving not only in packaging material but in labour.

Expanded polystyrene boxes have also been widely used for the transit of fresh fish. The advantages are obvious. Expanded polystyrene has very much better heat insulation properties than wood and thus less crushed ice has to be used to maintain the fish at a low temperature. It also obviates any damage from splinters. The clinching argument is cost. In some countries expanded polystyrene already competes with wood in price, and in the rest the gap is narrowing. In many instances the lightness in weight of the material has been exploited to give cost savings due to reductions in air freight. For example, Iceland uses an expanded polystyrene box of 25 kg capacity for the export of fresh salmon by air to the UK while scallops are exported from the UK to Paris, also by air. A reduction in transit time by the use of air transport makes a tremendous difference to the quality of the fish – aided by the insulation properties of the expanded polystyrene – but is only really economic if lightweight packaging is used.

Other developments include the use of open framework boxes, moulded from high-impact polystyrene, with folded fibreboard inserts that help to stiffen the structure and contain the produce. Boxes of this sort were first marketed in Italy. Also from Italy came the idea of a returnable, collapsible crate, moulded from polypropylene. The sides of the crate are hinged at the base so that they can be stored flat on the return journey with consequent savings in freight charges. They were first used for carriage of citrus fruits from Sicily to the markets in Rome.

The injection moulding industry has now produced a whole range of one-trip re-usable containers of all shapes and sizes, many of which are standardised to EC dimensions in order to fit defined palletised operations. Such containers have been designed to optimise the combination of lightness and strength, by utilising lattice, open-cage and ribbed structures. Sizes range from small produce trays, through Euro Containers of size 600 mm × 400 mm, right through to 0.5 tonne, 1200 mm × 1200 mm bulk containers, large transit cradles and full size pallets.

Polypropylene and HDPE are the two polymers competing in this market. Polypropylene offers better long-term creep and load-bearing (i.e. stacking) properties and higher temperature resistance (thus allowing hot cleaning or sterilisation without deformation). It also gives products with good definition, relatively free from strain and stress cracking, which is important in products of complicated design. HDPE, on the other hand, offers advantages in low-temperature and impact properties.

## 17.12 Modified atmosphere packaging (MAP)

Modified atmosphere packaging is described as a separate item because it is relevant to a range of products. MAP is essentially a method of packaging whereby the product inside the pack is surrounded by a modified gaseous atmosphere. The air is evacuated from the package and is replaced by a mixture of gases, the composition of which depends on the nature of the product.

As an example of MAP we can re-examine the packaging of fresh meat, discussed earlier in this chapter (see secton 17.9). One problem mentioned there was maintaining the bright red colour that the consumer associates with freshness. Another problem is bacterial growth. The use of MAP can deal with both of these problems by incorporating an oxygen-enriched atmosphere plus carbon dioxide which retards the growth of bacteria. One gaseous mixture that has been used successfully is 80% oxygen and 20% carbon dioxide. For pork, which is high in fat and so needs less oxygen, a successful mixture is 69% oxygen, 11% nitrogen and 20% carbon dioxide. The operation is a thermoform/fill/seal one, using trays made from a PVC/polyethylene laminate with lidding material made from a polyethylene/polyester laminate coated with PVDC. This gives a combination of good oxygen barrier and heat-sealing properties. Gas composition is not the only factor in MAP. To obtain the full benefits, temperature must be controlled all through the distribution chain, while good hygiene is also essential.

Other produdts that are benefiting from MAP are fish, cooked meats, baked goods, cheese, fresh produce and prepared salads.

# CHAPTER 18

# Pharmaceutical and health care packaging

The term 'pharmaceuticals' covers those products that are produced for the welfare of humans and animals. It includes medical substances administered orally, by injection or externally (in the form of a spray, cream or mousse). Health care items are devices used in medical or surgical procedures and include a wide range of products, from cotton wool swabs to tongue depressors and from scalpels to dialysis filters.

The correct packaging of pharmaceuticals and health care items is obviously important because any failure can lead not only to a dissatisfied customer but also to possible loss of life. Like foodstuffs, pharmaceuticals are often sensitive to external influences but the adverse effects of such influences on foodstuffs are usually though not always) observable by virtue of off-flavours, off-odours or changes in colour. Pharmaceuticals, however, may often be rendered useless for their purpose in a manner that is indiscernible other than by chemical, biological or other analytical methods. In such a condition they may be directly or indirectly harmful and dangerous to health.

Although the packaging of pharmaceuticals is cost sensitive, particularly for ethical (prescription) products, the main criteria are protection and function. Product assessment is an important prerequisite, therefore, to the development of any pharmaceutical packaging and the resultant package choice will be subject to a rigid programme of testing.

In addition to problems relating to protection of the product, it is essential that the package does not interact with its contents. Another problem that faces a pharmaceutical package is the fact that the product may have to be sterilised inside the container. This could entail the use of elevated temperatures, radiation or gases (such as ethylene oxide) and a material that is suitable from other aspects may well fail in this regard. Finally, the package may have to function as a dispenser (e.g. eye drops, ear drops, nasal spray, etc.).

A more recent problem with the packaging of pharmaceutical products is that of rendering the package child-resistant. A number of ingenious closures have been developed for use on bottles, jars or aluminium tubes. However, many of the really effective designs are also difficult to open by elderly or arthritic patients. The development of child-resistant closures would have been impossible without the use of plastics but this subject will be returned to later (in Appendix B).

In general, plastics are now widely used in the fields of pharmaceutical and health care packaging in spite of the problems outlined above. In fact, plastics are gradually becoming the first choice of packaging material for many pharmaceutical products. This choice of plastic, together with its ultimate acceptance for the product, involves the following steps:

(a) selection of the best type of plastic for the purpose,
(b) selection of a plastic grade which has food grade clearance,
(c) checking the selected material for residues, processing aids, additives and any other constituent,
(d) carrying out extractive tests using a selected simulant and checking the extractive for chemical substances and biological activity (e.g. toxicity or irritancy),
(e) carrying out accelerated storage tests using the product and plastic,
(f) carrying out formal stability tests for the full shelf-life of the product,
(g) performing actual or simulated user tests,
(h) purchasing production material to an agreed specification,
(i) checking all materials (packaging and product) against specifications (including the finished pack), and
(j) monitoring all complaints and any adverse reactions.

The ways in which plastics have satisfied the stringent requirements of pharmaceuticals packaging are probably best examined under the various product forms, i.e. solids, pastes, liquids, etc.

## 18.1 Solids

Solid pharmaceutical products fall into four main categories, namely:

(a) mixed powders for oral administration or local application, e.g. dusting powders
(b) effervescent granules, taken dissolved in water
(c) single-dose forms for oral administration, including tablets, pills, capsules (hard and soft) and lozenges
(d) single-dose forms for administration other than oral, including suppositories and pessaries

As far as powders are concerned, most are packed in flexible packaging of one sort or another (such as sachets) and so do not concern us here. Some powders are freeze-dried preparations and must be packed in hermetically sealed containers such as ampoules or rubber-capped glass bottles. Here, again, plastics have not yet gained a foothold.

In the packaging of effervescent granules, the most important requirement is the exclusion of moisture vapour. For consumer packs, glass bottles or plastics film or laminate sachets are widely used but plastic bottles have also been used, usually polypropylene or HDPE.

Coming now to single-dose forms for oral administration, there is a much wider choice for package forms. The choice depends on a number of factors, including the following:

(a) the amount of protection necessary
(b) product/pack compatibility
(c) cleanliness – low level of microbial and particulate matter contamination
(d) presentation – particularly for those products that may be the subject of impulse buying such as over-the-counter (OTC) items; presentation is also important as a means of inspiring user/patient confidence
(e) customer convenience – weight, size, ease of opening and reclosing, legibility of printing
(f) the method of filling
(g) cost

Bottles or tubes, fitted with screw caps or snap-on closures, are the commonest package forms. Glass bottles have been used for many years and have become the standard against which other containers are measured. Many tablets are packaged in polystyrene tubes but these do not provide sufficient protection for hygroscopic tablets, particularly for export to tropical countries. Polypropylene and HDPE are superior in this respect and have been used for tubes and for bottles.

One thing to be noted when looking at cleanliness is that plastic bottles can be produced 'clean' rather than being cleaned, by washing, after production. This entails the use of filtered air for the blow moulding, a filtered air area for the machine, the minimum of handling and clean packaging.

Whether plastics or aluminium tubes are used, it is necessary to minimise the movement of the tablets. This may be achieved by the use of cotton wool or other stuffing but there are now a number of plastics closures available with suitable internal fittings.

Suppositories and pessaries are moulded solid products, usually made from glyco-gelatine (with added medicaments). Consumer packs often consist of greaseproofed, fibreboard containers but rigid, thermoformed plastic trays have also been used.

## 18.2  Semi-solids

Semi-solids comprise those substances and preparations that are too thick or viscous to be treated as liquids but are not dry in nature. They can be subdivided into ointments, creams and pastes.

### 18.2.1  Ointments

These are usually made with greasy bases in which the required medicaments are dissolved, mixed or dispersed. They are usually anhydrous and the bases from which they are made include vegetable oils, mineral oils and greases. Most B.P. ointments can be packaged in plastics pots but volatile ingredients (such as oil of wintergreen) may interact with plastics (acting as plasticisers) or may permeate through the plastic.

One success story for plastics pots is the packaging of petroleum jelly. With glass pots there was the occasional danger of glass splinters in the product due to impacts during high-speed filling. If a glass splinter was then accidently rubbed into a baby's skin there would be loss of customer goodwill, even if the actual injury was not dangerous. The use of polystyrene pots has removed this possible danger.

Plastics jars, in general, are lighter, less breakable (require less protection) and more elegant than glass. In addition, plastics jars can be printed by silk screen and other methods, hence the need for paper labels is eliminated. This is an advantage because labels can become soiled in use, particularly by greasy products.

One interesting use of plastics is for the packaging of eye ointments. These were once always packed in metal collapsible tubes. The use of collapsible tubes is obligatory for those ointments (such as eye ointments) which are required to be sterile. These ointments are filled aseptically into previously sterilised tubes. Sterilisation of metal tubes is by steam or dry heat. Eye ointments are now packed in PVC tubes. This is because metal tubes may contain metal particles that could injure the eye. PVC tubes are not necessarily free from particles but any that are present are not so liable to cause eye damage. PVC tubes, of course, are not able to stand up to heat sterilisation methods so they are supplied already sterilised by radiation, in sealed polyethylene bags, ready to be aseptically filled.

### 18.2.2  Creams

These are usually emulsified systems and can be either oil-in-water or water-in-oil. Plastics pots (polystyrene or polypropylene) are again used for reasons similar to those outlined earlier.

## 18.2.3 Pastes

Pastes are normally for skin application and are usually stiff preparations containing a high proportion of powdered solids mixed in an oily or greasy base, with glycerine and water or with a gelatinous base. Plastics pots are used for paste products but the choice of glass, metal or plastics will depend on the quantities involved and the nature of the base and of the active principles.

## 18.3 Liquids

Pharmaceutical liquids are, in general, fluids in which are dissolved or suspended therapeutically active ingredients. The liquids used include water, alcohol, glycerine, oils or organic solvents such as ether, acetone, chloroform, etc.

Before looking at some of the individual pharmaceutical liquids it is worth noting one area where plastics have almost completely replaced glass. This is for medium bulk quantities where 5 gallon demijohns in either earthenware or glass and 10 gallon glass carboys were commonly used. Polyethylene containers have now replaced them to such an extent that 10 gallon glass carboys are no longer manufactured in the UK. The 5 gallon/25 litre and 10 gallon/50 litre polyethylene containers are particularly popular for export purposes because glass carboys are unacceptable to most shipping lines and the polyethylene containers are much easier to handle than glass carboys in iron baskets stuffed with wood wool. In spite of this there are a few liquid products for which neither plastics containers nor metal drums are suitable and glass is the only choice. These, relatively small quantities, are imported.

### 18.3.1 Aqueous oral preparations

Aqueous oral preparations include extracts, syrups, elixirs and linctuses. Extracts are prepared by extracting the active principles from crude drugs by percolation or maceration with water or dilute alcohol. Syrups are concentrated aqueous solutions of sugar. To these are added either medicaments (e.g. syrup of codeine phosphate) or flavouring agents, for use in the compounding of other preparations. Elixirs and linctuses are also pleasantly flavoured vehicles for potent medicaments, the tastes of which require masking. They may contain syrup, glycerine and alcohol.

Glass containers are still the main choice but plastics are making headway. The particular plastic used depends on the ingredients present, such as ether, chloroform or aromatic oils.

### 18.3.2 Aqueous non-oral preparations

These include lotions and eye drops. Lotions are usually intended for application to the skin or scalp and may be simple aqueous solutions (such as zinc sulphate lotion) or suspensions in water (calamine lotion) or alcohol (salicyclic acid lotion). The use of plastics bottles is still a minor one because of the nature of some of the ingredients. Phenolic substances, for example, may be present as in disinfectants or bactericides and be lost by permeation through the more common plastics such as polyethylene or polypropylene. Organic solvents are sometimes added to retain the active principles in solution and these, too, could be lost through permeation.

Eye drops are solutions of medicaments which are dropped into the eye. Like eye ointments, mentioned earlier, they must be sterile. The containers must, therefore, be sterilised before or after filling. The particular method of sterilisation and filling is dictated by the nature of the active ingredients present. When glass bottles are used the application of the solution to the eye is carried out using a glass pipette fitted with a rubber teat.

Plastics (usually polyethylene) bottles are often used as alternatives to glass ones because of their flexibility. Such containers are fitted with nozzles from which the contents can be dispensed by squeezing. They have the advantage that they minimise the risk of contamination by bacteria, introduced by the replacement of a pipette, which may itself have become contaminated by contact with an infected eyelid. Storage tests must be carried out, however, because certain of the bactericides used may be absorbed by polyethylene.

### 18.3.3 Injections

The vehicle most commonly used for injections is water which must, of course, be pure. Medicaments may be injected intravenously, intramuscularly, subcutaneously or intracutaneously but in all cases the injections must be sterilised and their containers must also be sterile and capable of maintaining their sterility until required for use. At one time, glass was the only material used for vials, ampoules, etc., but HDPE, polypropylene and PVC have all now been used. Plastics containers have one advantage over glass as they contain no alkali that could raise the pH of the injection solutions. Careful tests must be carried out to check that the polymer grade used contains no toxic additives capable of being leached out into the solution.

The form/fill/seal bottle system, described in Chapter 11, has been used for the sterile packaging of transfusion liquids. One variation has a moulded-on tab at the base of the bottle and this is used to hang the bottle at the patient's bedside. Another recent development incorporates a rubber septum.

### 18.3.4 Emulsions

Emulsions are mixtures of oil and water, either of which can be the continuous phase, the other being dispersed as droplets. Emulsifying agents are added to stabilise the emulsion. Plastics bottles have been used successfully for many emulsions but thorough testing is essential. An interesting example is that of Turpentine Linament BPC which is an oil-in-water emulsion. In spite of the fact that the turpentine is dispersed as droplets, with water as the continuous phase, the turpentine is still capable of causing environmental stress cracking of polyethylene. The problem is solved by the use of polypropylene or PVC.

### 18.3.5 Non-aqueous liquids

The commonest types of non-aqueous liquids are ethyl alcohol, other organic solvents and oils. Ethyl alcohol is widely used because of its solvent effect on many medicaments that are insoluble in water. In addition, it is non-toxic, palatable and has preservative properties. As mentioned earlier, it is used in the preparation of extracts. Other preparations based on alcohol are known as tinctures and 'spirits'. The latter are strong alcoholic solutions, often of flavouring agents, such as spirit of peppermint.

Although ethyl alcohol itself can safely be packaged in plastics containers, it is the other ingredients that must be considered. In the preparation of spirit of peppermint, for example, the oil of peppermint would soon permeate through the walls of a polyethylene bottle. Polypropylene, PVC or PET (or PETG) may be suitable but tests must always be carried out.

Other organic solvents used in pharmaceuticals include ether, chloroform, acetone and benzene. Preparations containing any of these have to be carefully tested before attempting to package them in plastics.

Oils used in pharmaceuticals are usually vegetable oils. One mineral oil that is fairly widely used is liquid paraffin and this is often packaged in HDPE. Olive oil, castor oil, almond oil, etc., can all be packaged in plastics. PVC is the most resistant but HDPE and polypropylene have also been used.

## 18.4 Health care packaging

In most instances, the devices involved in health care packaging must be delivered to the ultimate used in a sterile condition. The package plays an important role in maintaining such sterility up to the final point of use. As mentioned earlier, the range of health care items is very wide but there are basically only five main types of packaging used. These are: flat pouches, vent bags, gussetted pouches, pre-formed trays and lids, and thermoform/fill/seal packs. Only the last two come under the headings of rigid or semi-rigid containers so only these will be considered here.

Pre-formed trays and lids are more expensive than the pouches and vent bags but they do offer better protection and can be used to contain several items at a time, such as all the items needed for a single surgical procedure. Trays can also be made nestable so as to reduce storage space when empty. Trays are made from a range of polymers, including polyester (PET), poly (acrylonitrile), polycarbonate and acrylic multipolymer. The lids are usually made from a heat-seal-coated paper or from spun-bonded, non-woven materials such as Tyvek (trade name of Du Pont Co.). Spun-bonded non-wovens can be produced from a variety of polymers, including PET, nylon, polyethylene and polypropylene. Tyvek itself is based on HDPE.

Heat-seal conditions, especially heat and pressure, are often critical to the maintenance of sterility. Too high a seal pressure, for example, can force the molten heat-seal coating into the lid material and so produce a 'starved' joint. On the other hand, too high a temperature may cause distortion of the tray flange with a consequent reduction in the tray area.

One interesting example is the packaging of a knee implant in a blister package. Sheet made from PETG is vacuum formed into blisters, die cut, cleaned and then packaged into plastic bags and corrugated cartons. At the filling plant, the blisters are cleaned by air blowing and an antistatic solution is applied so that any loose fibre can be removed by wiping. The knee implant is first placed in an inner tray blister and sealed with Tyvek. This inner tray is then placed in an outer tray blister and the entire package is sealed with Tyvek to give safe and effective handling of the implant before and during surgery. PETG is used as the tray material because of its clarity and its ability to withstand sterilisation with gamma radiation.

The thermoform/fill/seal type of package can be filled at high speed and is, therefore, very suitable for large-volume items. Two rolls of material are involved in the operation. The lower layer consists of a thermoplastic laminate or coextrusion and is thermoformed to give a tray or blister. This is filled (either manually or automatically). The upper (lidding) layer is then applied and the finished packages are separated and packed into cartons. Because of the high cost of thermoform/fill/seal machines, production runs also need to be high (of the order of a million per annum of one type and size). Materials used include PVC, unoriented polyester and poly(acrylonitrile).

One of the important factors in material choice is the method of sterilisation employed. PVC is excellent from the points of view of clarity and ease of forming but is not easily sterilised. Steam sterilisation cannot be used because of PVC's heat sensitivity. Normal grades of PVC are also affected by radiation and must be stabilised which adds to the cost. Irradiation usually causes loss of HCl with subsequent structural changes in the molecule, leading to discolouration. Radiation-resistant PVCs are available which inhibit cross-linking and show low discolouration. The gas sterilisation process, using ethylene oxide, also has disadvantages for PVC. PVC absorbs ethylene oxide and a lengthy aeration cycle is often necessary to ensure a gas-free package.

Unoriented polyester sheet also thermoforms well and has good clarity. It is suitable for sterilisation by ethylene oxide and by irradiation.

Poly(acrylonitrile) is similar in some respects to PVC but is more resistant to radiation. It is also suitable for gas sterilisation. It is used in both blister packs and for thermoform/fill/seal applications.

One other material that has sometimes been used in both types of pack is polycarbonate. It is expensive, however, and its use is limited to specific applications where its high clarity, high temperature resistance and high impact strength are of particular interest.

Whichever package system is used there are problems in the opening and closing of the packs. The closure must provide a hermetic seal and this seal must be maintained during sterilisation, distribution and storage of the packs. On the other hand, many such packages are opened under emergency conditions where rapid access to the contents is required. These contents are also required to be presented in a sterile condition.

Cutting of the package is not the answer, even if a sterile cutting agent is used, because the outside of the package is contaminated and this contamination could be transferred to the contents. The usual solution is the use of peelable seals but this introduces fresh problems. A weak seal may be desirable for easy access but this can lead to seal failures during sterilisation and distribution, particularly if the size and shape of the contents are such that they exert an appreciable pressure on the seals. The provision of, say, a corner start will enable the peel strength to be increased while still providing rapid access.

## 18.5 The future

The difficulties surrounding the packaging of pharmaceuticals outlined at the beginning of this chapter have had a big effect on packaging material choice. For many years, glass and metals were the predominant materials of choice but today metals (other than for aerosols, aluminium foil and metallised films) have lost out significantly to plastics. Glass, too, is increasingly under threat from plastics.

The reasons for this growth in the use of plastics, particularly in the field of rigid and semi-rigid containers, are (a) the development of PET stretch blow moulded bottles and (b) the increasingly wide range of properties made possible by coextruded bottles, especially with the advent of high barrier materials such as EVOH.

The influence of PET can be seen from figures for PET usage. In 1988, the West European PET market for pharmaceutical packaging was estimated as 1000 tonnes while the projected market for 1995 is 5000 tonnes. PET has the important properties of crystal clarity combined with a shatter-resistance superior to that of glass.

Another advantage of plastics that will further increase their growth is their ability to produce containers by form/fill/seal processes, both by thermoforming and by bottle blowing techniques.

The final factor that enhances the usefulness of plastics in pharmaceutical packaging is design flexibility. The subject of child-resistant packaging has already been mentioned, but the other side of the coin is packaging for the elderly and those with limited dexterity. One example of the way in which plastics can assist here is the HDPE container known as the Jaysquare (trade name of Johnsen and Jorgensen Ltd). With its square shape and rounded corners it is easy yet comfortable to grip. Furthermore, it is fitted with an injection moulded closure, designed with a prominent groove in the top into which a lever can be inserted to give increased turning power. The closure is also heavily knurled to facilitate easy gripping, while the neck of the bottle has been modified so that the cap can be removed in three-quarters of a turn. A final feature is that the cap is moulded in bright orange to aid the partially sighted.

# CHAPTER 19

## Cosmetics and toiletries

---

Cosmetics and toiletries are not such a clear-cut group of products as the ones we have previously considered, namely, food and pharmaceuticals. The distinction between cosmetics and toiletries is also difficult to define. It has been said that cosmetics are used for adornment while toiletries are used for hygiene but there are many examples where the dividing line is crossed. However, the distinction is not important and it is convenient to discuss them under a single heading.

The packaging of cosmetics and toiletries is different from that of many other products, inasmuch as sales appeal usually plays such an important part. This does not mean that package cost is not a factor but it is not as important as in the packaging of household products or industrial items. Other packaging criteria, such as protection and function are also important. Protection is particularly important in the case of perfumery cosmetics or toiletries because the loss of the perfume, or of even one of its constituents, can lead to a totally unacceptable product as far as the consumer is concerned.

The cosmetics industry has been built on the idea of selling the promise of beauty and personal attractiveness so that the appearance and smell of the product must be in line with that promise. Any deterioration, such as off-odours, loss of odour, corrosion, caking of powders or drying out of creams and pastes may, therefore, have a disproportionate effect on sales

Product assessment, therefore, is again an essential prerequisite for effective packaging. In the case of cosmetics and toiletries it is particularly important to carry out this assessment of the requirements of the product as early as possible, since one may then be able to avoid certain problems by reformulation of the product.

Because the cosmetics industry is selling a promise (as well as the actual product) the package has a very big part to play in sales. Plastics are widely used (where other factors allow) because they can be moulded into a wide variety of attractive shapes and are available in an enormous range of colours. Polystyrene, in particular, is widely used because in addition to a huge range of shades, it can be obtained in a variety of pearlescent and metallic finishes. The appearance of plastics can often be further enhanced by the use of hot foil printing. Polystyrene containers for talcum powder were, in fact, one of the first rigid plastic containers to be used in the cosmetics industry – around about 1960. They are still used today and are injection blow moulded. In addition to the attractive range of colours available, polystyrene is popular because of its very smooth, glossy surface and the wide range of shapes that can be produced compared with tinplate or fibreboard. Although polystyrene is not a good moisture vapour barrier, it is completely satisfactory for talcum powder, which is not highly hygroscopic.

## 19.1  Packaging criteria

A brief look at the criteria of main concern in the packaging of cosmetics and toiletries will serve to illustrate the role of plastics in this field.

### 19.1.1  Appearance

In the light of the main requirements of cosmetics and toiletries we can begin to assess the suitability of plastics for their packaging. In general, as pointed out earlier, they satisfy the requirements of appearance (or sales appeal) to a very high degree. In addition to the range of colours and shapes, some plastics, such as PVC and PET, can be used to produce containers of sparkling clarity.

### 19.1.2  Protection

The main protection required in the packaging of cosmetics and toiletries is against perfume loss. Glass and metal containers provide complete barriers against perfume loss, so long as the closure is adequate, but plastics do not and this must be taken into account. Of course, the permeability of a plastic container depends on thickness so that adequate protection can often be obtained at a price. Low-density polyethylene is very permeable to the essential oil components of a perfume but high-density polyethylene and polypropylene are better in this respect. PET is even better and is likely to be more widely used in the future.

### 19.1.3 Function

In the field of cosmetics and toiletries the package is often called upon to perform some particular function, in addition to its job of containing the product. Lipsticks and eye make-up sticks are good examples of functional cosmetic packs while roll-on deodorants are good examples of functional toiletry ·packs. Here again, plastics score because of their versatility of design.

As with pharmaceuticals, the ways in which plastics have satisfied the requirements of cosmetics and toiletries packaging can be examined under the various product forms, such as solids, pastes, creams and liquids.

## 19.2 Solids

Solid cosmetics include such items as lipstick, make-up, stick deodorants and solid fragrances. Although we must consider these products as solids, they are, in fact, very soft solids with surfaces that are easily marred and, in stick form, are easily broken.

The packaging for a lipstick is quite complicated and consists of a godet (the small cup which holds the lipstick), a base, a sleeve, a barrel and a cover. The godet is usually injection moulded from polystyrene and has two lugs which are used in the twist-up mechanism. The base is also made from a plastic, usually polystyrene. It is a cylinder, closed at one end, and runs the whole length of the case. The godet (with the lipstick) fits loosely inside the base which has two vertical slots through which the godet lugs project. These lugs engage into two helical grooves inside the sleeve. The barrel is a tight fit around the sleeve and hides the twist-up mechanism from view. When the barrel is held and the base is turned, the lugs on the godet are forced to slide up or down the helical grooves in the sleeve, thus raising or lowering the lipstick. The final component is the cover which is a deep skirted decorative cap. This fits over the barrel and is removed prior to using the lipstick. Once again, polystyrene is the usual material.

Lipstick cases were once always made from metal, usually brass, but plastics have made considerable inroads. Even when a metal case is still used, the mechanism is usually of plastic. The efficient functioning of a lipstick case is very much dependent on correct dimensions and tolerances. The cover must fit tightly inside the barrel and the godet must slide smoothly up and down as the base is turned. Polystyrene is the ideal material because it is easily injection moulded and gives rigid components that are dimensionally accurate and stable. It is also, as we have seen before, ideally suited in terms of appearance.

The packaging for stick deodorants and anti-perspirants is simpler than that for lipsticks because a deodorant stick is stronger and of greater diameter so that it is unlikely to break in use. Deodorant sticks are not a luxury item and are never used in public so they do not need to be glamorously packaged. The usual pack consists of a body, screw cap and godet. The body, which is open at the bottom, is injection moulded from polystyrene while the cap and godet are injection moulded from polyethylene. The body is thick walled to protect against permeation. The godet is a sliding fit and is pushed up for use. An effort to upgrade the image of deodorant/anti-perspirant sticks *vis-à-vis* sprays and roll-ons has resulted in a twist-up pack with a polystyrene barrel and a screw-on cap of polypropylene. The polystyrene barrel is printed in four colours.

Solid fragrances are similar but the package is usually more decorative and elegant than for deodorants. A twist-up mechanism is used, similar to that used for lipsticks.

Pressed powders, as used for face and eye make-up, also come under the heading of solids. These are usually sold in fairly inexpensive compacts which are discarded when empty. The compacts are injection moulded from polystyrene in two parts, with a hinge, or from polypropylene in one part, utilising the ability of this material to form an integral hinge. Eye make-up compacts are similarly moulded and are usually rectangular in shape. The use of polypropylene integrally hinged containers means that only one moulding has to be produced and inventoried. It also saves on assembly costs.

Finally, we should not overlook the wide variety of thermoformed packs where the 'solid' product is a bottle, jar, lipstick case, razor, etc. These range from simple formings to take, say, a matching perfume and talc, to a large pack of men's toiletries, including a comb and a razor. Other formings have been used for false eye-lashes and nails. One very popular use for thermoformed trays is as gift packs, particularly at Christmas. PVC and polystyrene are the most widely used materials and further variations are available by the use of embossing, flocking, self-colouring and vacuum metallisation. Vacuum metallising, incidentally, is also used on lipstick cases, to give an up-market impression.

## 19.3 Powders and granules

One of the most common cosmetic powders is talcum powder, and the packaging of this was, to some extent, covered earlier in the chapter. An additional type of package sometimes used for talcum powder is the squeezable polyethylene 'puffer pack'. Although the puffer pack is very successful for many powder products, it has not proved so successful for talcum powders because of their perfume content which renders the powder somewhat sticky. Talcum powder is also widely packaged in polypropylene containers.

The other product under this heading is bath salts. These used to be packed in large glass jars but blow moulded polyethylene bottles are often used because of their light weight and resistance to breakage. They are also less likely to damage the bath or wash basin if they are dropped into it!

Both bath salts and talcum powder are commonly stored in the bathroom where the twin risks are water vapour and free water. Polyethylene is a barrier against both while polystyrene is a barrier against free water. As explained earlier, a high water vapour barrier is not usually required for talcum powder.

## 19.4 Pastes and creams

The list of cosmetics and toiletries manufactured in the form of pastes or creams is a very long one and includes toothpaste, face cream, hand cream, barrier creams, shaving cream, sun-tan cream, depilatories, foundation creams and hair creams or mousses. Most of these are packed in collapsible tubes. The original tubes were manufactured from tin or lead but tin is expensive and most metal collapsible tubes are now made from aluminium.

Plastic tubes were developed in the mid-1950s and were made from PVC or polyethylene. Later, polypropylene was also used. One advantage of plastic tubes is the fact that corrosion is no longer a problem. They are also free from the problem of fracture during use with a consequent loss of product through breakage of the tube. The disadvantage of the early plastic tubes was that they did not remain collapsed after squeezing but sprung back into shape when the pressure was released. When this happens, cream is drawn back into the tube from the nozzle and air is sucked in. This is a particular disadvantage in the case of thick creams or pastes because the air forms bubbles which prevent the dispensing of a smooth ribbon from the tube. This is not a problem with thin creams because the air does not form permanent bubbles but rises to the surface. This process is usually aided by the use of a full diameter cap on which the tube can stand upright, so that the air in the tube rises to the heat-sealed end, away from the nozzle.

Plastic tubes are not impermeable so that volatile ingredients can be lost through the tube wall. Permeation is not rapid, however, because of the wall thicknesses involved. External lacquers, often based on epoxy resins, can also be applied to improve the barrier properties and enhance the surface gloss.

The latest development in plastic tubes is the laminate tube, which combines the advantages of plastics and metal tubes. Laminate structures vary but all contain aluminium foil (to give the requisite barrier properties) and often contain paper (for stability). A typical structure would be:

- Outer layer – LDPE (transparent)
- 2nd layer – printed, white pigmented LDPE
- 3rd layer – paper
- 4th layer – LDPE
- 5th layer – an acrylic adhesive
- 6th layer – aluminium foil
- 7th layer – an acrylic adhesive
- Inner layer – LDPE

Printing is sometimes achieved by reverse printing on the outer polyethylene layer. Sealing of the laminate into a tube is carried out by radio-frequency energy which induces heat in the aluminium foil. Laminate tubes do not spring back to shape after squeezing so that there is no tendency to draw back air. Unlike metal tubes, however, they cannot be rolled up from the bottom so that in use they are flattened and retain their full length. The printing remains visible throughout use (unlike opaque PVC or polyethylene tubes). The biggest market for collapsible tubes is toothpaste, being about 250 million per annum in the UK. Laminate tubes now hold the major share of this market and are beginning to take over the packaging of other products such as cream shampoo.

Pastes and creams are also packaged in screw-capped pots or jars. Many creams were originally marketed in thick-walled opal glass pots and when high-impact polystyrene was suggested as a replacement, the thin-wall plastic pots were so much smaller that customers believed that they were receiving less product for their money. The answer was to produce a double-walled pot, with an air space in between, to match the outer size of the opal glass pots. Unfortunately, this answer to the problem was denounced as 'deceptive packaging', although both the external dimensions and the internal capacity matched those of the opal glass jars. The original plastic pots were injection moulded but many are now injection blow moulded. One of the earliest of these injection blow moulded pots was used for a leading brand of petroleum jelly. The pots were manufactured from basic (crystal) polystyrene and had excellent clarity. The main attraction of the plastic pots was the absence of the risk of chipping during high-speed filling (with consequent presence of glass splinters in the jelly).

The latest innovation in plastic pots and jars is the use of wide-mouth PET containers, produced by injection stretch blow moulding. These containers were developed for the food industry but were also found to be very suitable for cosmetic creams. Plastic containers, fitted with mechanical pump-action dispensers, are also providing an attractive alternative to the collapsible tube for toothpastes.

## 19.5 Liquids

Cosmetic and toiletry liquids include deodorants, anti-perspirants, suntan oils and milks, shampoos, hair conditioners and hair colorants, after-shave lotions, splash-on liquids and foam bath products.

HDPE and PVC bottles have been used for many years but PET is gaining ground. PET is said to be particularly suitable for alcohol-based toiletries. such as after-shaves, colognes and fragrances.

Shampoos, being detergent based, can cause environmental stress cracking in polyethylene containers and high molecular weight grades must be used in such instances. This stress cracking does not occur with either PVC or PET. Stretch blow moulded polypropylene is another candidate here. Not only is it not subject to stress cracking but it possesses contact clarity and is relatively low in cost.

Many cosmetic and toiletry liquids are dispensed from aerosol containers, usually glass, tinplate or aluminium. The use of plastics for aerosol containers has been tried many times over the years but no large markets have yet been developed. One of the earliest attempts was in the early 1960s. The plastic used was polypropylene and the container was injection blow moulded which gave the necessary control of wall thickness for a pressurised container. Polypropylene also gave the requisite barrier properties for the product (a range of colognes) and the low-pressure propellants that were used.

This was a one-off experiment. Higher pressure aerosols, like hair sprays, need a stronger container and much work was carried out using a polymer called poly(acetal) but without commercial success. The latest material to be put forward for use as an aerosol container is PET and some success has been achieved.

PET has the necessary barrier properties against gases and water vapour and is chemically inert to a wide range of products. Containers are manufactured by the injection stretch blow moulding process, the biaxial orientation so produced resulting in containers with a high tensile strength and excellent pressure resistance. The injection moulded neck finish produces the engineering tolerances that are necessary for accurate valve crimping.

The benefits of PET over tinplate include light weight, seamless construction, freedom from rusting and denting, all-round graphic potential and the availability of a variety of surface finishes. PET aerosols have been made to withstand pressures of up to 20 bar, which is 6 bar in excess of that required by BS 5597. The market for PET aerosols was put at 500 tonnes in 1988 and was projected to reach 5000 tonnes by 1995. Current indications, however, are that this growth rate was over-optimistic as the early promise has not been realised. Air-powered spray or aerosol dispensers are a more promising development, particularly for PET.

Incidentally, the market for other PET cosmetics and toiletries containers was estimated at 2000 tonnes in 1988 with a figure of 8000 tonnes predicted for 1995.

# CHAPTER 20

# Household chemicals

---

Household chemicals include items such as bleach, washing-up liquids, fabric softeners, multisurface cleaners and a range of detergents used in washing machines and dish washers. Plastics containers now play a large part in these markets. Far from being substitute materials, plastics are usually first choice when the packaging of newly developed products is considered. Two examples, here, are fabric softeners and carpet fresheners.

Factors that are important in the packaging of many household chemicals are protection, cost and function. Appearance (sales appeal) may also be important but not to the same extent as for cosmetics and toiletries. The flexibility of plastics design is widely utilised in the packaging of household chemicals. Moulded, integral handles, for example, give a better grip for containers holding dangerous chemicals, while trigger sprays and pump action containers facilitate the dispensing of a range of products.

As with other product types, the choice of container is often dictated by product form. Applications will be examined, therefore, according to the product form (solid, liquid, paste, etc.).

## 20.1 Solids

The commonest solid household product is washing powder (or granules) but this is usually packaged in fibreboard cartons. However, even here, there is a new plastics application in the form of a ball which is used to carry the appropriate amount of granules in intimate contact with the clothes being washed.

A similar product is the detergent used in dish washers. This, too, has long been packed in fibreboard cartons but now at least one manufacturer is using a high-density polyethylene container for its 3 kg size. The container has an integral handle, a pourer spout and a squeeze and turn screw cap (as a child-resistant closure).

Another household chemical sold in powder form is carpet freshener. This is sprinkled over a carpet, left for a few minutes, then vacuumed up. This is one of the products, mentioned earlier, that has been packed in plastics (usually HDPE) since its inception.

## 20.2 Liquids

There is a very wide range of liquid chemicals used in the home and the majority of these are packed in plastic containers. The largest group by far is washing-up liquids. In the UK they are packed in medium-density polyethylene containers whereas in the USA, high-density polyethylene rigid containers are used. The reason for this is interesting. In the USA and in the UK, glass bottles were originally used but an unbreakable container was sought by the detergent manufacturers. In the USA, tinplate containers were developed and these displaced glass bottles as the container of choice. When plastic bottles were developed they were made in the image of the (rigid) tinplate container (including the use of a cap which doubled as a measure). In the UK, meanwhile, work was also being carried out on developing a tinplate detergent container, but before this was commercially available, a composite plastic/metal container appeared on the market. This consisted of an extruded low-density polyethylene tube with metal ends. One end was fitted with a plastic spout and the liquid detergent was dispensed by squeezing the container wall. When plastic bottles were developed in the UK they, too, imitated the existing container and were made from LDPE to give a squeezable dispensing action. A medium-density polyethylene is now used, giving equivalent performance with a thinner-walled (and hence cheaper) bottle. The original LDPE tube with metal ends is still in use as a 'puffer pack' for various powder dispensers.

Plastics (mostly HDPE) have also established themselves in the fields of floor cleaners and floor treatments. Many of these bottles are made with integral handles. They are easier to hold than a plain bottle and also add to the sales appeal. The fact that plastic bottles are easily obtained in a range of colours enables the manufacturer to market a family of products in the same-shaped container but differentiated by means of colour. This has been done successfully in the case of an own-brand range of polishes for wood floors, vinyl floors and quarry tiles. Other cleaners packaged in plastic bottles include glass cleaners, tile cleaners and multisurface cleaners.

Two cleaning products packaged in PVC, rather than in HDPE, are Cusson's 1001 Carpet Cleaner and 1001 Foam Carpet Shampoo. The former uses amber-tinted, transparent PVC while the latter uses white pigmented PVC. The bottles are straight-sided, oval-shaped and with a moulded-in handy grip feature.

Liquid detergents for automatic washing machines are also on sale although powders/granules still constitute the largest section of this market.

We said earlier that fabric conditioners were one of the newly developed products that were packed in plastics from their inception. A design success for a British company selling bottles into Europe is an HDPE bottle with an integral dosing system. This has been used in Holland and Germany for the packaging of a concentrated fabric conditioner. The use of a concentrate means that weight savings are achieved, together with improved shelf-space utilisation. The integral dosing system (only possible with the use of plastics) enables the consumer to achieve the necessary control of the concentrated product.

Bleaches and lavatory cleaners are other products that have benefited from the use of plastics. Bleaches were originally packed in glass bottles but the risk of breakage, with such a hazardous product, facilitated the changeover to HDPE bottles when these became available. The original bleach bottles were simply copies of the glass ones but the trend for lavatory cleaners has been toward the use of bottles with shaped necks, to which are attached LDPE spouts. These then act as directional dispensers enabling the user to direct the product up under the top rim of the pan.

The field of household chemicals is a wide one and plastic containers have, by virtue of their design versatility, been found suitable for many of them. The usual material is HDPE. Examples include a range of liquid wax shoe polishes. HDPE is also used for the packaging of a bath stain remover. Product recognition is achieved by the use of a stocky, sloping shoulder container. Another HDPE container is a side-handled, white-pigmented flask for a cotton crisper. The container was said to have increased the sales of this product to such an extent that a normal year's sales were achieved in the first four months.

A very different pack is used for a nappy cleanser. The 2.5 kg size is packed in a pail-type container made from white polypropylene copolymer. The lid is made from LDPE. The pack has many reuse possibilities in the home and nursery.

A novel product, packed in an even more novel container, is a liquid formulation containing a germicide and a fungicide to kill bacteria, remove the source of bad odours, refresh and clean sports footwear. The polyethylene blow moulded bottles are plimsoll-shaped and the snap-off toe cap acts as a cover for an injection moulded polypropylene cap.

A large market for plastics is that of disinfectants. These, like bleaches, were once packaged in glass but their hazardous nature made them ready candidates for plastic containers. The usual material is PVC because of its clarity and chemical resistance.

The subject of aerosol containers was mentioned in Chapter 19. The problems of plastic aerosols have limited any use for household chemicals but sprays or jets can be obtained using 'trigger pumps'. These use plastic containers and dispensers. Because they use no propellants there are no problems of internal pressure, nor are products limited by product/propellant compatibility problems. The design of the dispenser can be varied to give a range of sprays or jets. A somewhat different design is used to dispense pastes or creams such as a hand cleaner.

# CHAPTER 21

# Miscellaneous products

---

Two large product areas that are conveniently dealt with in this chapter are motor oils and agricultural chemicals, but there are a number of other, smaller ones such as adhesives, greases, paints, automotive products and garden products that deserve mention.

## 21.1 Motor oils

The packaging of motor oils in plastics is one of the fields that has developed differently in various countries. The original method for retailing motor oils on service stations was by means of metal drums and dispensing cans. This was followed by reusable glass bottles but there were quality control problems with these resulting from their cleaning and the necessity to safeguard against substitution (instances were found where unscrupulous garage owners were putting used sump oil into branded containers).

Later, sealed metal (tinplate) containers were adopted which ensured that the motorist obtained the stated grade and measure. Such tinplate containers were capable of being filled at high speeds and any competing container had to have the same capability. In addition, any plastics container had to satisfy the following criteria:

1. It must have adequate chemical resistance.
2. It must prevent loss of the oil by permeation.
3. It must not affect the properties of the oil.
4. It must not contain any additives that can be extracted by the oil.
5. It must possess adequate rigidity and impact resistance.
6. It must provide protection for the contents against all environments likely to be encountered during distribution and storage.

From all these requirements was developed the so-called plastic 'fruit can', first in Germany and then in Holland and quite a few places outside of Europe. This container is blow moulded from HDPE by the false neck process. The unwanted false neck is then trimmed, giving a profile to which a metal lid is crimped. This type of container can be filled on lines designed for metal cans, with little or no modification, and is equally tamper-evident. A later modification in Germany reduced the cost by eliminating the tinplate lid and replacing it with polyethylene coated aluminium foil. The aperture was reduced but was still wide enough for fast filling. The container resembled a bullet in shape and was again blow moulded with a false neck. The trimming process was critical if a good surface was to be obtained for heat sealing the aluminium foil to the container. The size and shape of this bullet-type oil can made it rather difficult to 'up-end and leave' when filling the engine and an ingenious opening device was developed. This consisted essentially of a hollow spear. The end of the spear was used to puncture the foil while the hollow shaft was placed in the engine filler hole where it acted as a funnel.

In the UK, the use of plastics containers for motor oil has lagged behind that of the rest of Europe, mainly because tinplate prices were lower in this country. However, tinplate prices have risen faster than those of plastics over the years and there are now many examples of plastics motor oil containers on the market. The first plastics oil containers in the UK were in the larger (5 litre) sizes and in the area of speciality oils such as 2-T oil and outboard motor oils.

One of the problems in the smaller sizes (0.5–1 litre) was that of 'panelling' of the containers, particularly on outdoor exposure. It is caused by a combination of absorption of oxygen from the headspace, temperature fluctuations during storage, softening and swelling of the plastic and the use of high filling temperatures. The overall effect of these factors results in a vacuum being created inside the tightly sealed container, thus causing the walls to bend inwards at weak spots or where there are deformation stresses. Panelling can be counteracted by increasing the wall thickness, by using a higher density of polyethylene and by incorporating carefully designed strengthening features such as moulded-ribs. Vented closures are also sometimes used. These allow air to enter and compensate for the vacuum without allowing the contents to leak. Tall and narrow cylindrical containers tend to show the effects of panelling more readily than squatter ones. Shape also has an effect and panelling is less obvious in bottles of elliptical cross-section, compared with those of circular cross-section.

Larger (around 5 litre) sizes have been in use for some time and there have been many developments over the years. One is the incorporation of easy-pour (no glug) necks. Such a feature is particularly useful when aiming at some rather inaccessible engine filler hole. In the field of decoration, screen printing has been joined by the use of therimage (heat transfer labelling) and reverse-printed PVC sleeves. Stackability has also been developed, a particularly ingenious solution being a

Norwegian jerrican type of container which is stacked two by two, each pair of containers being turned through 90° in relation to the pair below.

There have also been developments in tamper-evident closures for all sizes of container. Two types will be described here. In one, a deep-skirted snap-on plastic cap is forced over a bead at the base of the can neck. The skirt is scored in such a way that pulling a tab removes a strip of plastic almost all the way round the skirt. The result is a shallow-skirted snap-on cap attached to a ring of plastic below the bend on the neck. The cap is thus a tamper-evident and a captive cap. The other type has a series of ratchet-like teeth moulded into the base of the neck. The cap has a matching set of teeth attached by means of thin spokes. The cap is screwed down until the ratchet teeth are fully engaged. When the cap is unscrewed, the ring of ratchet teeth is torn from the cap, thus giving evidence of opening.

## 21.2 Greases

Plastic tubes have been used for certain specialised greases since they avoid the difficulties with metallic contamination that are sometimes associated with the use of aluminium tubes. PVC, HDPE and LDPE have all been tried at one time or another but HDPE has been found to be satisfactory in most instances. For certain aviation greases, polypropylene is preferred for maximum resistance to the migration, from the grease, of specialised additives. Polypropylene has also been used for a range of automotive grease cartridges used in grease guns. One reason is the quality of the decoration possible while another is the fact that the rigidity of polypropylene is a great advantage during the high-speed filling operation.

## 21.3 Petroleum

Normal HDPE jerricans as made for the packaging of detergents are not suitable for the packaging of petroleum because of permeation. Thicker-walled containers have been made that are suitable, an example being a German one weighing about 750 g and with an overall average wall thickness of about 3.65 mm. The screw cap is well engineered and is provided with a rubber gasket. It has been claimed that at 20 °C the weight loss would be of the order of 7% per annum. An experimental container was once produced in Australia using nylon, but it was too expensive for commercial use. The latest technique is the fluoridation of HDPE containers (see section 21.10).

## 21.4  Paraffin

Problems associated with paraffin are similar to those encountered with the packaging of petroleum (though not to the same degree). Thick-walled containers are necessary to prevent significant permeation losses. The main advantage of an HDPE container for such applications is the elimination of corrosion hazards and consequent leakage during domestic use, which is sometimes encountered with tinplate containers.

## 21.5  Anti-freeze and brake fluids

Anti-freeze and brake fluids are satisfactorily packed in HDPE. Here, again, the lack of corrosion and contamination that is obtainable with plastics containers is of considerable marketing value. Tests have also shown that HDPE has no deleterious effects on the special additives incorporated in these products.

## 21.6  Car care products

There is a wide range of care care products on the market including anti-rust, radiator leakage preventives, radiator flush and bodywork repair products. All of these have been packed in HDPE blow moulded containers. Because of the hazardous nature of these products, the containers are fitted with tamper-evident closures.

Car upholstery shampoos and car polishes are also widely packaged in HDPE containers.

## 21.7  Paints

One of the problems in the early days of emulsion paints was the fact that, being water-based, they caused corrosion of metal cans, particularly at the cut edge of the curled rim of the can body. Developments in the design of the paint cans has greatly reduced the problem but plastics containers have now captured this market because they are corrosion free. Plastics containers also perform better in drop tests.

Another boost for plastics has been the development of a solid emulsion paint that is used straight from a plastics tray and applied by a roller. The material used is a polypropylene copolymer and the tray incorporates an integral handle for convenient carrying-home from the store. The trays also have airtight snap-on lids which hold clipped-in cards with user instructions printed on the reverse side.

## 21.8 Agricultural chemicals

Agricultural chemicals constitute a wide range of products, including insecticides, herbicides and liquid fertilisers. A useful division can be made, however, into solvent-based and water-based formulations.

### 21.8.1 Solvent-based agrochemicals

Solvent-based preparations where low aromatic content hydrocarbons are used can be packaged in thick-walled HDPE containers. However, some formulations can cause slight stress cracking and either high molecular weight grades of HDPE or polypropylene would then be recommended.

Emulsifiable concentrates are usually based on highly aromatic solvents and the packaging of these in HDPE would need uneconomically thick-walled containers in order to prevent high permeation losses. PET is now being used for products where HDPE would be unsuitable. Another advantage of PET is clarity which is important where the product has to be mixed with water and it is necessary to see the level of the contents. One example of the use of PET is in the packaging of a herbicide for dock, nettle and thistle control. Sizes range from 250 ml up to 5 litre and, in this application, the bottles are amber tinted to minimise any degradation of the product by ultraviolet light.

### 21.8.2 Water-based agrochemicals

There is usually little difficulty in packaging water-based agrochemicals in HDPE containers. However, for both water- and solvent-based agrochemicals, if the products are classified as hazardous, the packs have to meet UN requirements. PET is particularly appropriate for these.

## 21.9 DIY products

DIY products can be roughly divided into solid materials (such as screws, hinges, electric drill accessories, knobs and many others) and liquid or paste products (such as paint strippers, adhesives, putty and similar materials).

The first class of materials are packed either in polyethylene bags (with or without a header card) or in various thermoformed containers. The latter may be blister packs, using a backing card or all-plastic packs, often made on thermoform/fill/seal systems. The manufacture of these was described in Chapter 12. PVC is the most commonly used material although polycarbonate is sometimes used for the packaging of sharp objects. APET and polypropylene are also being used.

The packaging used for liquid or paste products is similar to that used for the other liquids dealt with earlier. HDPE is usually the preferred material, with PET being used for the more aggressive products.

## 21.10 Fluoridation

A fairly recent development which could lead to the wider use of HDPE is the in-line fluoridation of HDPE containers. As was seen earlier, one obstacle to the use of HDPE has been its permeability to non-polar solvents such as the hydrocarbons. The fluoridation process is based on the use of a dilute mixture of fluorine in nitrogen as a substitute for air during the bottle blowing process. The subsequent fluoridation causes a chemical modification of the HDPE surface which enhances its solvent barrier properties.

The mechanism for the penetration of a plastic by a solvent is a stepwise one that includes solvent wetting of the polymer surface, dissolution of the solvent into the polymer then evaporation of the solvent at the container wall. Because the chemical structure of HDPE is similar to that of many non-polar solvents, the container and the solvent are compatible, i.e. the solvent permeates the container walls via the preceding mechanism.

The fluoridation process depends on the formation of a fluorocarbon barrier layer on the inner surface of the bottle. The basic reaction that occurs with HDPE is a substitution of hydrogen atoms in the polymer chain by fluorine. The fluorocarbon barrier so formed, changes the surface characteristics of the polymer in terms of its polarity, surface tension and cohesive energy. These, in turn, have a major effect on reducing the wetting, dissolution and diffusion of non-polar solvents relative to HDPE. Fluoridation, therefore, is effective in reducing the permeability of non-polar solvents through HDPE. However, because in-line fluoridation modifies only those polymer molecules near the surface, there is no measurable change to the bulk properties such as tensile strength and impact resistance. Another related process for modifying the surface layer is sulphonation.

Finally, the technique of vacuum deposition of silicon dioxide (first used for films) has been adapted to the coating of bottles.

# PART 4

# PROPERTIES OF PLASTICS AND THEIR SIGNIFICANCE

CHAPTER 22

# Physical and chemical properties

## 22.1 General

The testing of plastics is usually carried out to determine their suitability for a particular application, for quality control purposes or to obtain a better understanding of their behaviour under various conditions. Where a new material has been developed, the manufacturer will need to be able to measure its performance with relation to other materials and so be in a position to assess the market likely to be available to him.

In general, standard test procedures will not provide the exact data necessary for a full prediction of operational behaviour. This is because, in most instances, standard tests for plastics materials were originally developed for quality control purposes. In such tests great emphasis is laid on consistency of test specimens, conditions and procedures and, where possible, on simplicity and reliability of test equipment. This is to ensure, as far as is possible, that results are reproducible between different laboratories.

In this chapter, the various properties of plastics mentioned in earlier chapters will be discussed in conjunction with the test methods normally used for their measurement. Some of the difficulties encountered when attempting to correlate test results with behaviour in use will also be examined. Some idea of the general difficulties can be obtained by considering the compromises that may be necessary for a particular application. In many instances the choice of material may depend upon a balance of stiffness, toughness, processability and price. For a given range of grades of a particular polymer, the rigidity increases as the impact strength decreases. Also, processing requirements may well place an upper or a lower limit on the molecular weight of the polymer that can be used and this will often influence the polymer properties to a great extent. Added to this is the fact that no single value can be placed on the stiffness or toughness of a plastic because:

1. the stiffness will vary with time, applied stress and temperature;
2. the toughness is influenced by the design and size of the moulding, the design of the mould, processing conditions and the temperature of use; and
3. both stiffness and toughness can be affected by environmental effects such as thermal and oxidation ageing, ultraviolet ageing and chemical attack.

Much of the early measurement of plastics properties used methods of test that had been developed for testing metals, textiles or wood. These were often unsatisfactory because of the differences in behaviour of the newer materials and it was not until some understanding had been obtained of the nature of plastics, and of their behaviour under conditions of stress at various temperatures, that real progress was made in standardising test methods.

### 22.1.1 Conditioning of test specimens

Many of the properties of plastics vary according to the ambient temperature or humidity or both and it is essential, therefore, to condition specimens before testing and to carry out tests under standard conditions. The effect of temperature is obvious in the case of thermoplastics; properties such as yield stress, for example, decreasing as the temperature rises towards the softening point of the plastic.

The effect of humidity can be a factor with certain plastics such as nylon and cellulose acetate. In conditions of high humidity, water is absorbed, with consequent impairment of physical and electrical properties, although these are restored to their original values when excess moisture is removed. Conditioning procedures are laid down in ASTM D.618-61.

### 22.1.2 Rate of testing

Because of the long chain structure of thermoplastics, their mechanical properties are time dependent. This is well illustrated by their behaviour under tensile stress. At very slow speeds the molecules can readily disentangle and the measured tensile strength depends mainly on the magnitude of the weakest intermolecular forces. At faster speeds there is little time for disentanglement or slipping and the breaking point does not occur until the strongest intermolecular forces have been overcome. It is difficult, therefore, to compare tensile test data from different sources unless the rate of clamp separation is standardised.

### 22.1.3  Fabrication of test specimens

The method of fabrication of test specimens can sometimes have an effect on the magnitude of the test properties. Orientation and rate of cooling (affecting spherulite size in crystalline polymers) can both vary according to the type of fabrication used and both can affect the test results. In connection with the above factors, and where relevant, a standard specification for testing a particular material usually lays down the method of preparing and conditioning a specimen before test. The number of test specimens is also stated in the relevant material specification but it may sometimes be desirable to alter this if a statistical approach is to be adopted.

In general, the physical testing of plastics can be classified as follows: dimensional, thermal, mechanical, transmission (permeability), electrical and optical tests. Of these, electrical properties are not intrinsically of interest in container manufacture and use, while most optical tests are aimed at measuring various film properties. Attention will be focused, therefore, on the first four types of test.

## 22.2  Dimensional tests

One of the commonest dimensional tests is the measurement of density or specific gravity and it is usually based on Archimedes' principle, i.e. the apparent loss in weight on immersion in water. Though basically simple, the preparation of the specimen, in the case of thermoplastics, has to be carefully controlled because the density of a polymer depends on its thermal history. An annealing treatment is, therefore, given in order that all specimens undergo the same heat treatment.

Density is usually quoted in grams per cubic centimetre and is thus numerically the same as specific gravity, which is defined as the weight of a given sample of material compared with the weight of an equal volume of water at the same temperature.

Specific gravity (or density) can sometimes be an important factor in cost. The price paid for plastics granules or powder is based on weight whereas the number of mouldings obtained is based on volume. Other things being equal, a slightly more expensive material with a lower specific gravity could, therefore, be cheaper in use.

## 22.3  Thermal tests

Among the thermal properties of plastics are thermal conductivity, coefficient of linear expansion, specific heat, softening point, heat distortion temperature and mould shrinkage.

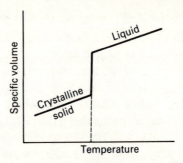

**Figure 22.1**  Melting curve of a low molecular weight crystalline solid. (Source: Miles and Briston, *Polymer Technology*, Chemical Publishing, New York.)

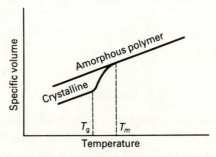

**Figure 22.2**  Melting curve of amorphous and crystalline polymers. (Source: Miles and Briston, *Polymer Technology*, Chemical Publishing, New York.)

Quoted properties for a plastic may sometimes include the melting point. This can be confusing if it is visualised as the same property as is exhibited by low molecular weight crystalline materials. When the latter type of compound is cooled from the molten state a sudden change occurs at its melting point as the material goes from the liquid state to a crystalline solid. The change is one from disorder to order and is accompanied by sudden changes in certain physical properties such as heat capacity and specific volume.

Figure 22.1 is a plot of specific volume against temperature for a low molecular weight crystalline material and shows the change at the melting point. Although similar changes occur when a molten thermoplastic is cooled, they are not so sudden.

Figure 22.2 is a similar plot for a crystalline thermoplastic and the more gradual change is clearly shown. The temperature at which the changes start to occur is known as the crystalline melting point and is denoted by the symbol $T_m$. The temperature at the second change in the curve is known as the glass transition point, or $T_g$. In general, thermoplastics above their glass transition point are rubbery in nature while below it they are hard and brittle. The upper line shows the behaviour of an amorphous polymer.

In actual use, of course, what we are interested in is some idea of the upper temperature of use. Can a particular container be hot filled, for example, or can it be heat sterilised after filling? Both softening point tests and heat distortion tests are designed to give the user of plastics some idea of safe temperatures in use but even here care must be taken when using the results of such tests because of the possibility of creep under continuous loads and, in the case of some polymers, an accelerated breakdown of the material due to heat. When the latter possibility exists, stabilisers are added by the plastics manufacturer and the efficiency of various stabilising systems is usually measured in terms of 'oven life'. The test is a simple one which involves placing samples of the plastic material in an air oven at 135 °C and examining them at intervals. Some of the more important softening points or heat distortion tests are described below.

### 22.3.1 Vicat softening point

The Vicat softening point test is defined in BS 2782: Part 1 and DIN 57302. Basically, it is a penetration-type test and results are expressed as the temperature at which an indentation of 1 mm occurs when the specimen is subjected to pressure by a standard needle (1 mm$^2$ surface area) under a standard load (usually 1 kg). The temperature is raised steadily during the test at a rate of 50 °C/h.

### 22.3.2 Deflection softening point

The deflection softening point test is also described in BS 2782: Part 1. The test specimen is clamped at one end and loaded at the other with a 20 kg weight. The assembly is then placed in an oil bath and the temperature raised at a rate of 1 °C/min. The temperatures at which angular deflections of 30° and 60° occur are noted and reported.

### 22.3.3 Heat distortion temperature

This is an American standard method (ASTM D.648-56) that gives some idea of the immediate tendency to deform under very low conditions of stress. Specimens 5 in. (127 mm) long, 0.5 in. (12.7 mm) wide and 0.25 in.

and 0.5 in. (6.35 and 12.7 mm) thick are supported 0.5 in. (12.7 mm) from each end and loaded at the centre to give fibre stresses of either 66 or 264 lb/in.$^2$ (46 400 or 185 600 kg/m$^2$). The temperature is then slowly increased (2 °C/min) until the centre of the specimen is deflected to the extent of 0.01 in. (0.254 mm). The temperature at which this deflection occurs at the two different loadings is then reported.

This test has certain limitations and the data obtained may be used to predict the behaviour of plastics at elevated temperatures only in applications in which the factors of fibre stress, temperature, method of loading and duration of stress are similar to those used in this test.

At the other end of the scale there are tests designed to evaluate the use of plastics at low temperatures.

### 22.3.4 Cold bend and cold flex temperatures

Among the tests concerned with the effects of temperature on the behaviour of plastics, BS 2782: Part 1 describes the cold bend and cold flex temperature tests. The former consists of winding a standard test piece around a mandrel 5 mm in diameter at a rate of 1 rpm. Tests are carried out at various temperatures, differing by 5 °C. The cold bend temperature is defined as the lowest temperature at which the specimens pass this test.

The cold flex temperature measures the temperature at which a standard angular deflection (200°) is obtained when subjected to a standard torque.

## 22.4 Mechanical tests

The mechanical properties of plastics vary over a wide range from soft and flexible to hard and brittle, with many intermediate combinations. It is, therefore, difficult in some instances to lay down standard tests for the measurement of a particular property that is suitable for all plastics. Some of the more important tests are described below.

### 22.4.1 Tensile properties

Tensile properties include tensile strength, elongation and Young's modulus. Two figures may be quoted for tensile strength, namely, ultimate tensile strength and yield strength. The relation between these two, and their relative importance, can best be seen by reference to a stress/strain curve. A typical stress/strain curve is shown in Fig. 22.3. Other properties that can be measured from a stress/strain curve are elongation (at yield and break) and the modulus of elasticity. The latter is given by the ratio of stress divided by strain and is a measure of the stiffness.

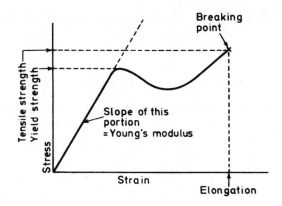

**Figure 22.3**   Typical stress/strain curve. (Reproduced from Briston, *Plastics Films*, 3rd edition, 1989, Longman, London.)

Before discussing tensile properties in detail it will be helpful to define some of the terms involved.

*Stress*

Stress is the ratio of the force exerted on a body to its cross-sectional area, i.e.

$$\text{Stress} = \frac{\text{Applied force}}{\text{Cross-sectional area}}$$

*Strain*

The strain is a measure of the change in dimensions of the test piece when a force is applied to it, and is calculated with reference to its original size, i.e.

$$\text{Strain} = \frac{\text{Total elongation}}{\text{Gauge length}}$$

*Gauge length*

Gauge length is the original length of that part of the test piece over which the change in length is determined.

### 22.4.2 *Ultimate tensile strength*

The ultimate tensile strength is the maximum tensile stress that a material can sustain, and is taken to be the maximum load exerted on the test specimen during the test, divided by the original cross-section of the specimen.

### 22.4.3  Yield strength

The yield strength is the tensile stress at which the first sign of a non-elastic deformation occurs and is the load at this point (known as the yield point) divided by the original cross-section of the specimen.

### 22.4.4  Elongation

Elongation is usually measured at the point where the test specimen breaks and is expressed as the percentage of change of the original gauge length.

### 22.4.5  Young's modulus

Young's modulus is the ratio of stress to strain over the range for which this ratio is constant, i.e. up to the yield point. It is a measure of the force necessary to deform the test piece by a given amount and so it is also a measure of the intrinsic stiffness of the material.

During the straight line portion of Fig. 22.3, the strain is proportional to the applied stress and when the stress is removed the test piece returns to its initial length. After the yield point, the strain is greater than it would be for a linear relationship and the non-proportional part of the strain is irreversible. In other words, a moulded article that has passed the yield point never returns to its original shape. This means that the yield stress is usually a more practical figure than the ultimate breaking stress except in cases of hard and brittle polymers where the yield point is ill-defined.

Although Fig. 22.3 is shown as a typical stress/strain curve there are other types of curve and these have been classified by T. S. Carswell and H. K. Nason (*Modern Plastics*, June 1944) as shown in Fig. 22.4.

The various curves represent the following types of polymers:

1. *Hard and tough*: such materials have a high modulus of elasticity, yield stress, ultimate tensile strength and elongation at break. Typical materials showing such a curve are cellulose acetate and certain of the polyamides.
2. *Hard and strong*: this type of material has a high modulus of elasticity, yield stress and ultimate tensile strength. It also has a moderate elongation at break. Typical polymers giving this type of curve are rigid PVC formulations and polystyrene copolymers.

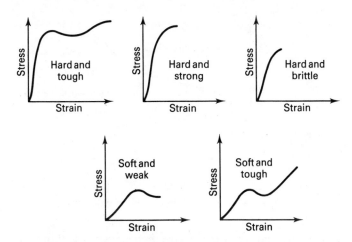

**Figure 22.4** Typical stress/strain curves obtained with polymers.

3. *Hard and brittle*: the hard, brittle type of material is characterised by a high modulus of elasticity and ultimate tensile strength. There is no well-defined yield point and the elongation at break is low. Typical examples are polystyrene, polymethyl methacrylate and certain phenol formaldehyde polymers.
4. *Soft and weak*: such materials have a low modulus of elasticity, yield stress and ultimate strength. The elongation at break is moderate. This type of behaviour is characteristic of soft polymer gels.
5. *Soft and tough*: these materials have a low modulus of elasticity and yield point. The elongation at break is high and the ultimate tensile strength is usually much higher than the yield stress. Examples of materials giving this type of curve include rubber and plasticised PVC.

A great deal of information about a material can be obtained from the shape of its stress/strain curve. In addition to the numerical values for tensile stress, Young's modulus, elongation, etc., it is possible to obtain some idea of the toughness of the material by measuring the area under the curve. This area is a measure of the energy needed to break the test specimen and hence is directly related to toughness.

### 22.4.6 Creep

As mentioned earlier, the speed at which the test pieces are stressed is important, especially for thermoplastics, since their mechanical properties are time-dependent by virtue of their long chain structure. There are many types of tensile testing machines available but the basic principle of them all is that a film strip is held at one end by a fixed clamp and at the other

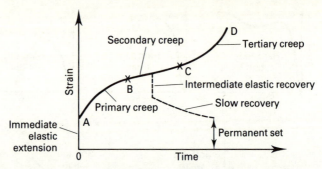

**Figure 22.5**  Typical creep curve for thermoplastics. (Source: Miles and Briston, *Polymer Technology*, Chemical Publishing, New York.)

end by a movable one. The clamps are then drawn apart and the tension in the strip is measured at various clamp separation distances. Tensile testing machines are usually fitted with multispeed gear boxes so that several speeds are available on the one instrument. Most machines are also fitted with charts and the stress/strain characteristics of the material under test are recorded automatically. Computer systems are also available which can be used in conjunction with tensile testers to give print-outs and visual display unit read-outs.

One effect of the time dependence of mechanical properties is that where long-term stresses are concerned a plastics article may rupture or deform by an unacceptable amount at a constant stress well below its short-term yield stress. This continuous deformation with time, under constant load, is known as creep. It is also encountered with metals, though to a lesser extent. Creep can be studied by applying a constant load to a standard test specimen and measuring the strain at intervals over long periods of time. The resultant values for strain are plotted against time when a curve of the type shown in Fig. 22.5 is obtained. When the load is first applied there is an immediate elastic deformation (O–A in Fig. 22.5), which can be calculated from Young's modulus for the material.

During the initial stage the strain increases at a diminishing rate (section A–B in Fig. 22.5) and this is followed by secondary creep (B–C in Fig. 22.5) where the strain increases at a constant rate. The final stage is known as tertiary creep and (C–D) here the strain increases at an accelerating rate. This stage is often accompanied by crazing or cracking of the test specimen. Thermoplastics exhibit varying resistance to creep. Those materials with high intermolecular forces (e.g. hydrogen bonding) exhibit low creep, as do those thermoplastics that are below their glass transition points under the conditions of test.

In practice, it often happens that the load is applied intermittently and for limited periods. When the load is removed there is an immediate elastic recovery followed by a further slow recovery. The test specimen does not normally return to its original dimensions, there being a residual strain or permanent set (Fig. 22.5).

For meaningful results that can be used in practice, it is necessary to obtain a series of curves at different stress levels and at different test temperatures. Any increase in temperature increases the magnitude of the creep because of the reduction in the attractive forces between molecules.

Packaging applications where creep is likely to be a problem include crates (where long-term static stresses are encountered at the bottom of a stack) and bottles containing carbonated drinks (internal pressure).

### 22.4.7 Flexural strength

Flexural strength is sometimes referred to as cross-breaking strength or the modulus of rupture. It is the maximum stress developed in the surface of a prescribed shaped test piece when supported near the end and loaded at the centre until failure occurs. Flexural strength is quoted in $kN/m^2$ or $lb/in.^2$

### 22.4.8 Impact strength

The measurement of impact strength is probably the most controversial topic in the subject of plastics testing and many criticisms have been levelled against the methods used for its measurement. Unfortunately, no one has yet put forward a universally acceptable substitute so that tests such as the Izod or Charpy continue to be used in spite of their disadvantages. Indeed, although it may be true that the test data obtained are of doubtful quantitative significance, they are still useful as a basis for estimating the relative shock resistance (or, perhaps more correctly, the relative resistance to stress concentration) of the various plastics. Impact testing has become standardised around the Izod test, in which a notched specimen is clamped in a vice and is struck by a swinging pendulum. The energy lost by the pendulum in breaking the test piece is registered on a dial. A popular variation is the Charpy in which the test piece rests on two anvils (one on either side of the notch) and the pendulum strikes the test piece immediately behind the notch. The different methods are illustrated diagrammatically in Fig. 22.6.

With thermoplastic materials the test specimens may be either injection moulded or compression moulded. The method of preparing the specimens must be stated when test results are published, because this has an effect on the magnitude of the results.

One fault of the Izod test is that the energy lost by the pendulum consists not of the energy needed to break the specimen, but of a complex

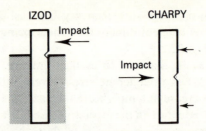

**Figure 22.6**   Diagrammatic representation of Izod and Charpy impact tests. (Source: Miles and Briston, *Polymer Technology*, Chemical Publishing, New York.)

figure consisting of the energy necessary to produce the first crack, the energy necessary to shear off the top half of the specimen and the kinetic energy in the broken-off piece which is manifested in its flight through the air after impact. Only the first figure represents the true impact strength and it has been maintained by some researchers that the remaining two factors can contribute to errors as high as 100% of the true impact strength.

Another criticism of the Izod test is that the results are influenced by the 'notch sensitivity' of the material under test. It will be seen that although the Charpy test overcomes the first objections to some extent, it is still liable to errors due to 'notch sensitivity'.

A low Izod value may indicate notch sensitivity but may straightforwardly be due to an intrinsically low impact strength. One way of determining whether a material is indeed notch sensitive is to carry out a series of impact tests with pieces having different notch tip radii. The Izod values are then plotted on a logarithmic scale against the notch tip radii on a linear scale. The steeper the slope of the curve, the greater is the notch sensitivity of the material and this gives clear warning to the designer to avoid sharp radii and other points of stress concentration in the moulding.

Another factor to be considered when looking at Izod test figures for any two materials, with a view to assessing their relative usefulness for moulding, is the question of stresses in the final article. Two polymers may have similar Izod impact strengths and yet behave very differently in practice. This can be due to differences in flow. The one with the highest flow gives a more shock-resistant article, especially if the moulding has thin walls or has complex flow paths.

An alternative method of measuring impact strength and one which simulates more closely the knocks likely to be encountered in practice, is the falling weight impact test (BS 2782). There are many variations in respect of the test pieces used but basically the test consists of dropping a known weight from a known height onto the test piece. The estimated level at which 50% of the samples are likely to fail is normally reported in this test.

Test pieces that have been used for this test include compression moulded sheets and injection moulded shallow dishes. The sheets are usually mounted on a circular support and the point of impact is applied at the centre. Injection moulded dishes (having their 'gate' at the centre of the dish) are inverted and located so that the point of impact is 20 mm away from the centre, the area near the gate being one of the recognised weak spots of an injection moulding.

In addition to the average level at which failures occur in the falling weight impact test, the type of failure can also be of importance in judging the relative impact performance of a material in practice. There are generally three types of failure, namely,

(a) tough (or ductile) failure in which the material yields and flows at the point of impact, producing a hemispherical depression which, at sufficiently high impact energies, eventually tears through the complete thickness;
(b) brittle failure in which the specimen shatters or cracks through its complete thickness with no visible signs of any yielding having taken place prior to the initiation of the fracture; and
(c) an intermediate or 'bructile' failure in which some yielding or cold flow of the specimen occurs at the point of impact, prior to the initiation and propagation of a brittle type crack (or cracks) through the complete thickness of the specimen.

If, in a given test, failures are all of the tough type, a reliable and reproducible measure of the impact strength of the sample can be obtained. If, however, differing types of failure are observed, a much wider variation in impact values may be obtained from repeat tests.

### 22.4.9 Hardness

Hardness is a complex property to measure and expresses the resistance to deformation. It is usually measured by some modification of the methods used in the testing of metals, the relevant standards being ASTM D.785-62 and ASTM D.676-55T. The basis of these tests is measurement of the indentation caused by the application of a steel ball or diamond cone under a standard load. Different sizes of penetrators are used for the different ranges of hardness and these are given letters which are quoted at the same time as the hardness number. Standard methods are the Rockwell hardness and Shore D tests for rigid materials; and BS softness number and Shore A tests for plasticised materials.

Taking the Rockwell test as an example, the procedure is as follows. A steel ball is applied to the test specimen under a minor load. This indents the surface slightly and ensures good contact. The gauge is then set at zero. The major load is then applied for 15 s and removed, leaving the minor load still in position. The indentation remaining is read directly off the dial.

The sizes of the balls used, and the loadings applied, vary and values obtained with one set cannot be correlated with values from another set.

Rockwell hardness can differentiate relative hardnesses of different types of a given plastic. However, since elastic recovery is involved as well as hardness, it is not valid to compare hardnesses of various kinds of plastic purely on the basis of this test.

## 22.5  Transmission tests (permeability)

The rate at which a gas or vapour will pass through a plastic membrane depends on a number of factors. These are of three types: those that are governed by the properties of the membrane, those that are dependent on the properties of the permeant (gas or vapour) and those that are concerned with the degree to which interaction may occur between the membrane and the permeant. The process by which the gas or vapour permeates through a plastic membrane is known as permeation and consists of solution (or absorption) at one surface of the film, followed by diffusion through the plastic (under the influence of a concentration gradient) and ending by desorption at the other surface. After a relatively short period of time a steady state will be reached and the gas or vapour will permeate through the plastic at a constant rate (if the pressure difference between the two sides is maintained).

The quantity of gas, $Q$, passing through the film is directly proportional to the difference in gas pressure on either side of the film, and inversely proportional to the thickness of the film. In addition, it is proportional to the time during which the permeation has been occurring and to the exposed area.

In other words:

$$Q \propto \frac{At(p_1 - p_2)}{x}$$

where $Q$ is the quantity of gas that passes through the film; $A$ is the surface area in contact with the gas; $t$ is time; $(p_1 - p_2)$ is the partial pressure differential; and $x$ is the thickness of plastic.

The foregoing expression can be put in the form of an equation, as shown below:

$$Q = \frac{PAt(p_1 - p_2)}{x}$$

where $P$ is a constant for a specific combination of gas and plastic at a given temperature. The factor $P$ is variously known as the permeability factor (or '$P$-factor'), permeability coefficient or permeability constant. It should be emphasised that the above equation applies only to steady-state conditions.

The position with regard to vapours, including water vapour, is less clear. For one thing, the steady-state condition is reached more slowly, while there may also be chemical interactions between the permeant and the plastic.

The practical importance of permeability hardly needs emphasising and a number of tests have been devised for its measurement.

### 22.5.1 Gas permeability

Measurement of gas permeability is carried out under controlled conditions of temperature and pressure. In the pressure increase method the film acts as a partition between a test cell and an evacuated manometer. The pressure across the film is normally maintained at one atmosphere. As the gas passes through the plastic, the height of the mercury in the capillary leg of the manometer changes. After a constant transmission rate is achieved, a plot of mercury height against time gives a straight line. The slope of this line can be used to calculate the gas transmission. Gas permeability measurements are described in ASTM D.1434 and BS 2782: Part 5: 1970.

Another method is that of concentration increase. Two gases are used, a reference gas and the test gas. A partial pressure difference across the barrier material with respect to the test gas is created without a difference in total pressure. Because the method of measuring the concentration of the test gas can be specific to that gas even in the presence of other gases or vapours, equipment can be developed in which the relative humidity of both the test and the reference gases can be controlled. This is not possible where equipment has to be evacuated and is very important when determinations are carried out on barrier materials having a hygroscopic base. Various methods of measuring the concentration change are used, including chemical analysis, gas chromatography and radioactive tracer techniques.

### 22.5.2 Water vapour permeability

One popular method for measuring the water vapour permeability of basic plastic materials is the dish method, described in BS 2782: Part 5: 1970.

The test specimens, usually thin films or sheets, are sealed over the mouths of metal dishes containing a desiccant, using wax as the sealant.

The dishes are weighted initially and then placed in a temperature and humidity controlled cabinet. Weighings are carried out at regular intervals and the gains in weight are measured. In one method the temperature of the cabinet is maintained at 25 °C $\pm$ 0.5 °C and the relative humidity at 75% $\pm$ 2%. These conditions are taken as representative of 'temperate' climates. For 'tropical' conditions the tests are performed at 38 °C $\pm$ 0.5 °C and 90% $\pm$ 2% relative humidity.

Where a quick indication of the potential barrier properties of a plastic is required, it can be clamped in an instrument where one side of the plastic is exposed to a high humidity atmosphere while the other side is in contact with dry air. As water vapour diffuses in, it is detected by an infra-red absorption cell or by a resistor, the value of which is affected by changes in relatively humidity. The instrument then measures the time required for a particular change in relative humidity as detected by the cell or resistor.

## 22.6 Tests on containers

All the physical tests described so far are carried out on moulded test pieces or, in the case of permeability, on film or sheet. There are, however, many tests that can be carried out on the finished container and these will often give results more in keeping with the end-use performance.

Before considering the methods used for testing various properties of the bottle, a little should be said about the production of standard bottles used for the tests. This is essential if any sort of comparison is to be made between the behaviour of bottles at various ambient conditions. The aim should be to produce a bottle that has a constant weight, possesses an acceptable variation in circumferential wall thickness, has been perfectly centred with respect to pinch-off, is free from contamination, 'fish-eyes' and severe 'melt fracture', and does not possess weakness at the weld, mould parting lines and pinch-off region. When optimum conditions relevant to each material grade and machine have been determined, the processing variables are standardised. Bottle weight and wall thickness are checked during processing and a visual inspection made for any faults in general appearance such as centring of base pinch-off, presence of 'fish-eyes', contamination, etc.

Returning to the subject of container tests, impact tests on plastics bottles may well be useful when a change in material and/or bottle design is contemplated and will quickly show up any fundamental weakness that may lead to leakage of the contents during transit. As with the Izod and other impact tests already described, bottle impact tests must be carried out under carefully controlled conditions if the results are to be of any value. The tests described have been developed with the aim of comparing different materials as well as assessing the influence of processing conditions on the impact properties of the bottle. Optimum processing conditions are essential, otherwise failure at weak positions on the bottle, such as weld and mould parting lines and pinch-off regions, will completely mask any material differences.

Two methods are of particular interest, namely, the 'Bruceton staircase' and percentage brittleness methods. In both instances stoppered 500 ml bottles, filled with water to give a total weight of 450 g, are dropped down an unplasticised PVC tube onto a level, rigid base plate. The bottles are conditioned in a test bath, either at 20 °C ± 1 °C or 0 °C ± 1 °C for at least half an hour, (at the latter temperature a little anti-freeze is added). Two tubes are used. One is 8 ft (2.44 m) long and is drilled to allow drop height variations of 6 in. (152 mm), while the other is 4 ft (1.22 m) long and is graduated for 3 in. (76 mm) variations in drop heights. In the 'staircase' method the impact energy is varied by altering the height of fall and determining the height at which half the bottles fracture. In the percentage brittleness method a large number of bottles are dropped from a fixed height.

### 22.6.1 Staircase procedure

The number of bottles required is 28. A bottle is rapidly transferred from the conditioning bath to the 8 ft (2.44 m) tube and is dropped from a height of 8 ft (2.44 m). If failure results, further bottles are tested, reducing the drop height by 1 ft (0.30 m) each time until no failure occurs. In this way the approximate failure height is established, using not more than eight bottles.

A bottle is then dropped from the height determined during the preliminary testing. If a failure occurs, the type of failure is recorded and the next bottle is dropped from a height 6 in. (152 mm) lower. If the bottle does not fail, then the drop height is increased by 6 in. (152 mm) for the next bottle. This procedure is continued for 20 bottles, increasing or decreasing the drop height by 6 in. (152 mm) according to whether the previous bottle withstood or failed the impact. For each bottle a record is made of the impact height and the result obtained, i.e. non-failure, brittle failure or tough failure. The position of the failure is also noted. If it is evident from the preliminary tests that the approximate failure height is considerably below 4 ft (1.22 m), e.g. around 2 ft (0.6 m), then the drop impact testing is carried out using the 4 ft (1.22 m) tube with standard increments or decrements of 3 in. (76 mm).

Let $O$ represent a non-failure and $F$ represent a failure. Then draw up

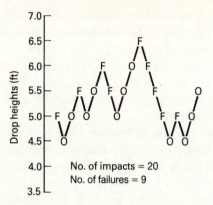

a column of drop heights on the left-hand side of the page and record the results of consecutive drops as they are obtained.

The calculation of impact strength (height to produce 50% failure) is as follows:

$$\text{Impact height} = \frac{(n_0 H_0 + n_1 H_1 + n_2 H_2 + \cdots n_i H_i)}{n_0 + n_1 + n_2 + \cdots n_i} \pm e \text{ ft}$$

where $n_0$ is the number of least frequent events at height $H_0$ ft; $n_1$ is the number of least frequent events at height $H_1$ ft; $n_i$ is the number of least frequent events at height $H_i$ ft; $e$ is the constant height increment between heights $H_0$, $H_1$, $H_2$, etc.

If the least frequent event is a non-failure, use plus $e$; if the least frequent event is a failure, use minus $e$.

In the example shown, the calculations are based on failure $(-e)$ as this was the least frequent event.

$$\text{Impact height} = \frac{(3 \times 5) + (3 \times 5.5) + (2 \times 6) + (1 \times 6.5)}{3 + 3 + 2 + 1} - 0.5 \text{ ft}$$

$$= \frac{15 + 16.5 + 12 + 6.5}{9} - 0.5 \text{ ft}$$

$$= \frac{50}{9} - 0.5 \text{ ft} = 5 \text{ ft}$$

### 22.6.2 Percentage brittleness procedure

The number of bottles required for this test is 50. From a knowledge of the results obtained from the staircase procedure, the impact height is determined that will give good differentiation between materials. From tests carried out on polypropylene bottles at 20 °C, a height of 8 ft (2.44 m) has been found suitable, while at 0 °C a height of 3 ft (0.9 m) gives suitable results.

### 22.6.3 Permeability

Permeability can also be investigated by using bottles, and one method for measuring weight loss will be described. The weight of a standard 500 ml bottle (of the particular type and grade of polymer being considered) plus the seal, is recorded and 450 ml of the product is introduced into it. Sealing can be carried out by means of a stainless steel insert incorporating an oil-resistant rubber O-ring which is placed into the bottle neck. A polypropylene cap is then screwed on at a constant torque of 2.26 N m (20 in. lb). The total weight of the assembly and contents is then recorded and the weight of the product in the bottle is determined. A total of six bottles is usually sufficient, three being stored in a constant temperature room at 23 °C while the other three are stored in an oven kept at 50 °C. The bottles stored at 23 °C are weighed at time intervals of 1, 2, 4, 8, 16, 32 and 52 weeks while those stored at 50 °C are weighed after 1, 2, 4, 7, 14, 21 and 28 days. The percentage loss in weight is calculated for each bottle, together with the mean percentage at each storage condition (ignoring any bottles that show an excessive weight loss due to faulty sealing). After each weighing, the seal between the O-ring and the bottle neck is checked by ascertaining that the screw caps are maintaining the required torque. The complete results (per cent weight loss against time) are plotted on linear–linear graph paper and the points joined to produce the best fitting line. Results at 23 °C and 50 °C are plotted on the same graph, where possible, so that a comparison can easily be drawn.

The situation with regard to the testing of plastics bottles is made more complex by the fact that the increase in permeability with temperature is not the same with all liquids, even in the case of chemically similar compounds. It follows, then, in the case of commercial products (which are usually fairly complex mixtures) that the change in composition due to permeation through the container walls at one temperature may bear no relation at all to the change at some other temperature. Table 22.1 illustrates the wide range in permeability/temperature relationships that may be encountered.

As can be seen from the results given in Table 22.1 the factor is not only widely different, even for a family of chemicals, but there is not even a regular change in the factor as one progresses through the series. Looked at in another way, we see that whereas at 20 °C methyl alcohol is lost at 2.1 times the rate of *n*-propyl alcohol, at 60 °C it is lost at 8.4 times the rate.

*Table 22.1* Increase in permeability rates with temperature (HDPE).

| Product | Weight loss (% per year) | | Factor |
|---|---|---|---|
| | 20 °C | 60 °C | |
| Methyl alcohol | 3.0 | 73.0 | 24.3 |
| Ethyl alcohol | 2.5 | 26.0 | 10.4 |
| *n*-Propyl alcohol | 1.4 | 8.7 | 6.2 |
| *n*-Butyl alcohol | 1.0 | 14.4 | 14.4 |

Oxygen permeability can also be measured on finished containers. The most widely used method is the MoCon oxygen-specific test equipment in which the permeating oxygen is catalytically combined with hydrogen in a fuel cell.

## 22.7 Chemical properties

The effect of chemicals on a plastic is obviously an important factor when assessing its suitability as a container for a particular product. The effect of a range of chemicals on particular plastics may often be obtained from the plastics manufacturer. The datum is normally obtained by subjecting a test specimen to the chemical under test, usually by total immersion under specified conditions. Any change in appearance or weight is noted at intervals, as is the change in a particular property such as tensile strength. As it is not possible to generalise about the effects of specific chemicals, tests should always be carried out where any doubt exists. This is particularly important in the case of mixtures even if information is available on the separate components. The action of chemicals can also be affected by internal stresses so that published results may sometimes differ from those obtained in practice.

### 22.7.1 Environmental stress cracking

When polyethylene is biaxially stressed in contact with certain environments such as soaps, detergents, wetting agents and silicone oils, failure by cracking can occur. The stresses may be impressed on the polymer by external forces (such as bending or twisting of a strip or the torque of a screw closure) or may be 'frozen-in' stresses caused by cooling of the polymer while in a state of stress as, for instance, at a badly designed pinch-off region. The resistance to environmental stress cracking of polyethylene varies greatly and is dependent on the nature and magnitude of the stress, the nature of the environment and on the thermal history of the polymer. As far as the polymer itself is concerned, the factor that influences the resistance to environmental stress cracking more than anything else is the molecular weight. The resistance to environmental stress cracking increases with increasing molecular weight but in practice there is an upper limit set by the processability of the polymer.

Investigating the environmental stress cracking properties of a product is important and much thought has been given to the question of suitable test methods. One method is based on the behaviour of a polyethylene strip under a standard stress. It can be adapted to the evaluation of different grades of polyethylene (by using a standard environment) or to the investigation of the stress cracking properties of different chemical environments (by using a grade of polymer having a known reaction when in contact with a standard environment). The method was devised by the Bell Laboratories in the USA and was later adopted as a standard by the ASTM. In essence, the test consists of taking a strip of polyethylene 1.5 in. long by 0.5 in. wide by 0.125 in. thick ($38 \times 12.7 \times 3.2$ mm) and scoring a line along the central 0.75 in. (19 mm) of the strip to a depth of 0.020–0.025 in. (0.51–0.64 mm). These test strips are then bent along their axes through 180° with the scored line on the outside. The strips are then inserted into a holder which maintains them in the bent position. The holder and strips are then inserted into a test-tube filled with the product under test and the test-tube kept at a temperature of 50 °C. The time to failure is noted and recorded. If a time of greater than 30 h is obtained in this test, it can be assumed that no stress cracking will occur in practice, provided that the bottles are produced under optimum processing conditions. If the strips fail in less than 30 h, then further tests must be carried out on actual bottles.

If it is desired to investigate the influence of bottle design or processing variables on the properties of the bottle, then the tests are carried out using a standard stress cracking medium. If, on the other hand, the stress cracking properties of a particular product are being investigated, then standard bottles are used. In either case the utmost care must be taken to produce the bottle under optimum processing conditions, otherwise the results will be worthless.

The bottles are partially filled with the product under test and then stored in an oven which is maintained at a temperature of 50 °C. Three bottles are stored upright while a further three are kept in the inverted position. This ensures that the regions of relatively high stress concentration (i.e. the neck and the base) are maintained in contact with the product. The bottles are inspected at intervals and bottle failures are noted. After long experience of this test, it is considered that if no failure occurs during 14 days' storage at 50 °C then no failure from stress cracking should be encountered at ambient temperatures for periods up to two to three years.

Practical points to be noted include the following;

(a) The amount of product put into each bottle should be standardised as indications exist that the volume of product in the bottle has an effect on the stress cracking life. In the case of the 500 ml standard bottle mentioned earlier, 150 ml of product is recommended.

(b) Bottles should be sealed with a stainless steel insert and rubber O-ring as detailed in section 22.6.3.

(c) The bottles should be placed in beakers large enough to contain the bottle contents should failure occur.

(d) The bottles should be examined within 1 h of the commencement of test to ensure that no leakage has occurred at the necks of the inverted bottles.

(e) Further examination should be carried out twice daily, i.e. early morning and late afternoon. Difficulties in determining the exact time of failure exist when such failures occur during the night, and in practice the most equitable solution is to deduct 8 h from the total time if failure is first observed at the early morning inspection, i.e. taking the mean time between the early morning and the previous afternoon times.

(f) If failure has apparently occurred in the inverted sample, a check is made on the seal at the O-ring. If it is considered that leakage has been due to faulty sealing and not stress cracking, the volume of liquid in the bottle is restored and the test continued. If leakage continues, a fresh bottle is substituted.

### 22.7.2 Panelling

When certain products are packed in plastic bottles and tightly sealed, collapse of the bottle walls can occur to a greater or lesser degree. This phenomenon is often referred to as 'panelling' and when severe can be so unsightly as to affect sales of the bottle and its contents. This wall collapse is caused by the creation of a vacuum inside the bottle. The creation of the vacuum may be caused by one or more of the following:

(a) loss of the product by permeation through the container wall,

(b) hot filling of the bottles and subsequent sealing while still hot,

(c) the absorption, by the product, of gas from the air space above the product; the gas most usually absorbed is oxygen,

(d) temperature fluctuations in storage which can cause differential gas permeation, or

(e) swelling of the container by absorption of the product by the plastic.

The effect of a given vacuum on any bottle is dependent on the strength of the bottle walls and on the design of the bottle. Anything that reduces the strength of the bottle walls will increase the incidence of panelling. This includes uneven wall thickness and softening of the bottle by certain products.

Observations of panelling can be carried out, at each weighing, on the bottles used for permeability determination. The degree of panelling can be assessed as follows:

1. mono-panelling – panelling on one side only (graded slight to severe)
2. di-panelling – panelling on two sides (graded slight to severe)
3. complete collapse

### 22.7.3 Staining or discolouration

Staining is another phenomenon that is looked for in the bottles used for permeability tests. It is most usually caused by absorption of the colouring matter in the product but may also be a sign that the product is causing chemical degradation of the polymer. The latter will usually be shown up by a comparison of a few fundamental physical properties measured before and after storage of standard test pieces in the product. Changes in tensile strength and elongation are two possible properties that can be checked.

### 22.7.4 Effects of the container on the product

Of equal importance to the effect of the product on the container is the reverse situation. One of the ways in which products can be changed is in colour and this is easy to see. Another is in odour and this is rather more difficult to detect, especially when the change is small. Even in cases where the change in odour is readily detectable, it is not easy to assess the change quantitatively, partly because of the subjectivity of odour assessment and partly because 'odour fatigue' in the assessors makes it difficult to check more than three or four samples at one time. A common problem encountered in this field is loss of odour from a perfumed product. While it is relatively easy to fix a point in the storage time when the product has lost all its odour, there are also cases where one or more components of a complex perfume (and most perfumes *are* complex) are lost preferentially so that the whole balance of the perfume is destroyed. This can be checked quickly by gas chromatography on the outside of a filled container held within a larger sealed vessel.

The problem of changes in flavour is closely related to that of changes in odour, and in both cases the only person who can give an answer to the question, 'Is the product still saleable?', is the manufacturer of the product.

Preferential absorption by the container or permeation through it, of one or more components of the product, may affect the product's performance in other ways. An emulsion, such as a furniture or car polish, may separate out into separate phases because of loss of the emulsifying agent or one of the phases and this could affect its efficiency as a polish. The simplest test here would be a comparative performance test with a similar sample of polish stored in glass. It is essential in the case of storage at elevated temperatures that the polish in the plastic bottle is checked against a sample stored in glass at the same temperature, since any deterioration in performance could be caused by the elevated temperature alone.

## 22.8 The importance of bottle design

After carrying out the tests described earlier, one may often be faced by results which, at first sight, seem to point to the fact that the product is unsuited for packaging in a plastic bottle. There may be a distinct tendency to panelling, for instance, or the permeability losses may appear to be too high. However, it is sometimes possible by close attention to bottle design to minimise such defects and produce a perfectly acceptable package.

Let us take, first, the question of panelling. Except in the case of product loss by permeation, none of the factors leading to panelling are intrinsically bad, and even the product loss may be acceptably low from other points of view and yet still cause panelling. We are faced, therefore, with the fact that if the panelling can be disguised or minimised in some way, a perfectly acceptable package may be produced.

The question of disguise is probably the more difficult one but it is not impossible. If wall collapse is not severe, the bottle can be so shaped that the collapse does not alter the shape of the bottle too drastically. Perhaps the worst possible shape for a bottle is a squat cylindrical one where even a slight vacuum can cause distortion to a quite noticeable degree. A better shape would be a bottle with an elliptical cross-section. Bottle design can also be made to minimise wall collapse simply by strengthening the container without necessarily using thicker walls. Waisted bottles are better than parallel-sided ones in this respect, while it is also possible to increase wall rigidity by moulding-in vertical or horizontal ridges.

Permeability, too, is a function of bottle design. Apart from the obvious solution of thickening the bottle walls, it should be noted that the percentage loss in weight is also a function of the volume/surface area ratio of the container. The shape with the least surface area for a given volume is the sphere so that one tends towards this shape as much as possible if permeability has to be reduced. This is of fairly limited value, of course, but may well be the deciding factor if the permeability loss is borderline. It is unfortunate that this conflicts with the requirements for the reduction of panelling but bottle design, as with most other things, is usually a matter of compromise.

Not surprisingly, bottle design can have a big effect on the resistance to impact (with the possible exception of LDPE bottles which can be produced in almost any shape without risk of breakage). Sharp corners should be avoided as should surface designs likely to cause local areas of stress concentration. The above remarks also apply to environmental stress cracking.

# Additives for plastics

---

Part of the BSI definition refers to their being solid, *composite* materials, i.e. there are materials present other than the basic polymer. This is because of the fact that although high molecular weight polymers are relatively stable compounds, the materials of commerce may embody imperfections that can affect their performance, either during processing or in use. Such imperfections may be in the structure of the polymer (the inclusion of other atoms, random and short chain branching, etc.) or in the presence of impurities such as catalyst residues, residual or unreacted reagents or low molecular weight polymers.

Additives may be necessary, therefore, to stabilise the plastic, assist its processing or improve its end-use performance. Some of the additives used are described in the following pages, together with an account of their functions.

## A.1 Processing aids

### A.1.1 Anti-blocking agents

Thin films tend to stick together, or 'block', especially if they have a smooth surface without irregularities. This can cause difficulties in handling and processing film, and is particularly annoying where packs of film are used for wrapping. Additives are used to give microscopic surface roughness and are usually very fine powders of low solubility in the plastic, which then migrate to the surface of the film. The most common anti-blocking additive is finely divided silica in the concentration range 0.1–0.6%. Anti-blocking agents may also sometimes be used in vending cups or tubs to aid de-nesting during filling operations.

### A.1.2 Antioxidants

Although plastics as a class are relatively stable compounds, many of them are susceptible to attack by oxygen, particularly at high temperatures.

The way in which additives can slow down or inhibit oxidation can be seen by looking at the oxidation processes that are usually involved. Briefly, it may be said that with most polymers the initial effect is either cross-linking or depolymerisation. With cross-linking the initial products are higher molecular weight, relatively infusible gels. These react further to give, eventually, compounds of lower molecular weight. Polyethylene is an example of this type of behaviour.

Depolymerisation occurs via chain scission, with the formation of lower molecular weight oxygenated compounds such as aldehydes, acids and hydroxy acids. Attack normally starts at double bonds already present in the molecule or at tertiary carbon atoms (thereby forming a double bond). Polypropylene is a typical example of this type of reaction, with the eventual formation of very low molecular weight products, such as butyric acid.

In both cases, as a rule, the reactions are chain reactions, often involving hydroperoxide radicals. The reactions can, therefore, be slowed down or inhibited by compounds that interrupt the chain reaction at some point.

Like all chemical reactions, the oxidation processes are highly temperature sensitive. In addition, many of them are also greatly affected by other factors such as catalyst residues, surface area and UV radiation. To a considerable extent, therefore, antioxidants cannot be considered in isolation and the stabilisation systems used in practice often combine to give optimum protection in relation to oxidation, temperature and UV radiation. In addition, antioxidants often show marked synergism, so that many antioxidant systems comprise two or more compounds, notably hindered phenols and organic sulphides. Hindered phenols, such as BHT (butylated hydroxy toluene), act as free-radical scavengers or as polymer chain terminators. Other types of antioxidants, such as organophosphites, act as hydroperoxide decomposers.

### A.1.3 Antistatics

All polymers are very good electrical insulators and will, therefore, retain electrostatic charges developed by friction between different layers or particles of the plastic itself, between the plastic and the processing equipment, or by electro-ionisation from dust, radiation or other sources. Such charges give rise to problems in processing and during end use.

In processing, the two main problems are: (a) adhesion, either between individual layers or particles of the plastic itself or between the plastic and processing machinery; and (b) discharging to earth by sparking. The latter can lead to pin holes in film, local overheating, electric shocks to operators and interference to machine operation generally. In extreme cases it can be a fire hazard.

To be effective an antistat must have two prime properties: (a) a reasonable degree of electrical conductivity and (b) a marked tendency to migrate to the surface of the film or container . The latter requirement is dictated by the concentration of the electrostatic charge itself at the surface and also by the fact that the only routes to earth for the charge lie over the surface.

Although many materials have been tried, the majority of those effective in actual use are either glycol derivatives or quaternary ammonium salts. Both tend to migrate to the surface but the conduction mechanisms differ. Glycol derivatives do have some intrinsic conductivity but their main effect is the sorption of moisture from the atmosphere (due to their hygroscopic nature) and it is this which acts as the electrostatic charge's path to earth. Quaternary ammonium compounds, on the other hand, are ionised to some extent in the anhydrous state and so act as conductors in their own right.

### A.1.4  Heat stabilisers

Most conversion processes use heat and, with some plastics, it is necessary to protect them from decomposition during their stay at a relatively high temperature. Some polymers such as LDPE, polyamides and condensation polymers generally, are stable enough not to require heat stabilisers.Polystyrene requires a small amount of stabilisation but PVC, HDPE and polypropylene may require significant quantities. The thermal decomposition of plastics follows a similar path to that of the decomposition caused by oxidation. Moreover, the two reactions are closely interlinked. Oxygen catalyses thermal decomposition while heat greatly accelerates oxidation.

As with antioxidants, there is a wide range of heat stabilisers available. The choice in any case is dependent on the actual temperatures likely to be encountered, the duration of protection required, the extent to which oxygen is present and the presence (or otherwise) of antioxidants. Where the finished container is to be used in food or pharmaceutical packaging, the additives must also be non-toxic.

The most common heat stabilisers are barium/cadmium, organotin and lead compounds. Calcium/zinc and antimony compounds are also used. An important class of compounds is that of secondary stabilisers of which the epoxy materials, such as epoxidised soy bean and linseed oil, are the most important. They have very little influence when used alone but when used in conjunction with other heat stabilisers they have a pronounced effect in improving heat stability.

### A.1.5  Lubricants

Molten plastics have extremely high viscosities and have highly non-Newtonian flow characteristics. It is sometimes necessary, therefore, to add materials that will reduce the frictional forces. Lubricants may be internal or external. Internal lubricants are those which reduce the melt viscosity and may be almost any compatible and stable compound. Most plasticisers (see later) will act as internal lubricants. Other substances used include paraffin oil, petroleum jelly and fatty acid esters and amides.

External lubricants reduce friction at the surface where the plastic melt contacts the processing equipment (e.g. the screw and walls of an extruder). External lubricants should not be too soluble in the plastic melt so that the concentration of the lubricant at the surface is increased. Metallic stearates such as calcium and zinc stearates are widely used.

### A.1.6 Mould release agents

In all moulding processes there is a possibility that the moulded object might stick to the mould. The problem is worst with injection moulding where pressures are high. In such cases mould release agents are applied to metal surfaces to prevent the sticking of the moulding during processing. The majority of mould release agents are silicones but calcium stearate is also used.

### A.1.7 Nucleating agents

Certain plastics may undergo deleterious changes, during heat processing, other than actual decomposition, the formation of undesirable crystallite structures being one such change. Polypropylene homopolymer, for example, is liable to form large crystals when heated and then cooled normally (i.e. relatively slowly, without quenching). This reduces the impact strength and the clarity of the product. The addition of nucleating agents encourages the formation of crystallites. Two types of nucleating agents are tertiary butyl benzoic acid or its salts, and salts of aromatic sulphonates. Nucleating agents are also used in the manufacture of expanded polystyrene sheet (see section A.2.3).

### A.1.8 Plasticisers

According to the International Union of Pure and Applied Chemistry, a plasticiser is a substance incorporated into a plastic or an elastomer to increase flexibility, workability and distensibility. It may act by reducing the melt viscosity (see also section A.1.5), by lowering the glass transition temperature ($T_g$) or by lowering the elastic modulus of the plastic. The mechanism by which plasticisers achieve their effect is by causing separation of the polymer chains, with a consequent reduction of the intermolecular forces.

Plasticisers are used for brittle polymers, such as PVC and cellulose acetate, to render them more flexible and to reduce their processing temperatures and so reduce the risk of thermal degradation. The proportion of plasticiser is usually quoted as parts 'per hundred resin' (phr) rather than per hundred total plastic (%). In other words, 10 phr means 10 parts plasticiser to 100 parts resin (= 10 parts plasticiser to 110 parts total plastic = 9.1%).

Plasticisers should be colourless and compatible with the resin. Because the plasticiser can only do its job while it remains in the resin, it must have a low volatility. Finally, for food contact uses, the plasticiser must be non-toxic and free from taint.

Phthalic acid esters are the most common plasticisers but maleates, adipates and citrates are also used. Low molecular weight polyesters and epoxidised oils are used as non-volatile plasticisers. Plasticisation by the incorporation of an additive is also known as external plasticisation. The separation of the polymer chains can also be achieved by incorporating a comonomer during polymerisation. This brings about the presence of side chains at intervals along the main polymer chain. This method of plasticisation is known as internal plasticisation. An example of internal plasticisation is the copolymerisation of small amounts of vinyl acetate with vinyl chloride.

### A.1.9  Slip additives

These are additives designed to reduce the coefficient of friction and are really only of importance in the case of films. In form/fill/seal machines, for example, thin films tend to 'drag' when pulled over metal such as a forming plate. Slip additives act as external lubricants, the most commonly used being fatty acid amides.

## A.2  End-use additives

### A.2.1  Antioxidants

As already mentioned, the main use of antioxidants is as a processing aid. However, some long-term oxidation does take place, particularly where relatively high temperatures are encountered in use (contact with hot foods, infra-red ovens, tropical areas, etc.). The principles and the materials used are similar to those described in section A.1.

### A.2.2  Antistats

These, again, have been mentioned in section A.1 but there are also end-use problems due to the pick-up of electrostatic charges. These are

(a) the adhesion of thin films (with possible sparking on separation) and (b) dust pick-up during storage or on retail shelves. The antistats used are the same as those already described.

A somewhat different aspect of the antistatic problem is that of packaging certain electronic devices and components. An increasing number of these are sensitive to electrostatic discharge (ESD) and electrical overstress. Ordinary antistats are not effective in producing 'static safe' containers and carbon black-loaded polymers are often used.

### A.2.3 Blowing agents

A number of plastics are capable of being formed into a foam but the one of greatest interest here is expanded polystyrene (EPS). Foaming may be carried out to provide heat insulation, increased protection against impact or increased stiffness.

EPS can be obtained either in moulded form or as sheet which is then thermoformed into cups, trays, etc. The foaming agent for moulded EPS is normally pentane. If the moulding is properly made and aired, most of the blowing agent should dissipate, leaving a residue below 0.5%

In EPS sheet formation the blowing agent is either pentane or a fluorinated hydrocarbon, depending on the method of manufacture. One method starts with polystyrene beads that have been impregnated, under pressure, with pentane. Nucleating agents (see section A.1), such as talc or a citric acid/sodium bicarbonate mixture are added to provide foaming sites and so produce a fine, uniform cell size. The other method utilises ordinary polystyrene beads and the gassing is carried out in the extruder. The blowing agent is usually a fluorinated hydrocarbon but can also be pentane. A nucleating agent is again necessary.

### A.2.4 Brighteners

Many plastics, in their natural form, do not have the transparency of glass or the white hue that is associated in the public's mind with purity and high quality. There are exceptions, such as polycarbonate, crystal poly-styrene and poly(methyl pentene) (TPX), but in general, plastics have an off-white or yellowish colour, coupled with translucency, rather than clarity. Brighteners are used, therefore, to render the material more attractive. They operate by absorbing incident radiation, converting it and emitting radiation of a higher frequency. The eye then registers this as a whitening or brightening effect.

Examples of brighteners are stilbene and thiophene derivatives in concentrations of about 0.01%. Their use in food contact applications is limited by virtue of possible toxicity but examples of their use may often be seen by cutting open a white polyethylene bottle and looking inside.

### A.2.5 Colourants

Colourants for plastics can be classified into two main categories, namely, pigments and dyestuffs. Pigments may be further subdivided into inorganic pigments and organic pigments.

Pigments are white, black, coloured, metallic or fluorescent solids that are insoluble in the medium into which they are dispersed. Particle size is small, normally in the range 0.01 to 1.0 μm in diameter. Pigments produce their colouring effect by the selective absorption of light but because they are solids they also scatter light, which is undesirable when transparency is required.

Organic pigments have high colour strength and brightness, low specific gravity, high oil and plasticiser absorption and are sensitive to heat and light. Some organic pigments have a very small particle size, so that light scattering is reduced and they act more like dyes. Inorganic pigments are more opaque, weaker in tint and are less bright. They are, however, more resistant to heat, light, chemical attack, weathering and migration (or leaching). Their specific gravities are greater than those of organic pigments and they are cheaper.

Dyestuffs may be defined as natural or synthetic organic chemicals that are soluble in most common solvents such as water and alcohol. They possess very good colour strength and are available in a wide colour range. The specific gravity of dyes is usually low and they are also normally characterised by low plasticiser absorption. However, they do possess some disadvantages, including poor heat stability and light resistance. Dyes also tend to bleed out from the polymer and rub off from the surface.

The requirements of a colourant can be summarised as follows:

(a) low specific gravity
(b) relatively low in cost
(c) stable to heat and light
(d) weather resistant
(e) good chemical resistance
(f) low oil and plasticiser absorption
(g) brightness and high colour strength
(h) easy dispersion
(i) solubility in suitable solvents

Selection of a colourant is also governed by the plastics to be coloured and the processing method employed. For example, PVC (and its copolymers) liberate hydrochloric acid at high temperatures so that acid-sensitive colourants such as ultramarine blue or cadmium red must be used with caution. Molten nylon acts as a strong acidic reducing agent and it can decolourise some dyestuffs and pigments. Colourants may also react with the nickel-bearing stabilisers used in some film-grade polypropylenes.

### A.2.6 Impact modifiers

Impact modifiers are added to PVC blow moulding compounds in order to overcome the brittle characteristics of unmodified PVC resins. Typical impact modifiers are acrylic resin derivatives, methyl methacrylate/butadiene/styrene terpolymers (MBS), chlorinated polyethylene, EVA and ABS. The final choice of modifier is not only dependent on the impact resistance properties required in the finished bottle. It also depends on considerations such as any loss of clarity, residual odour and, of course, cost.

### A.2.7 Slip additives

Most of the requirements for slip additives are in connection with processing (see section A.1) but there are certain end-use applications as, for example, in dispensing from sheaves of film.

### A.2.8 UV screening agents

UV radiation has a harmful effect on many products, in particular, certain foodstuffs. If these effects are likely to be significant they can be greatly reduced by the addition of UV screening agents to the container. Screening agents act by absorbing or reflecting the UV radiation. Reflected radiation is dispersed harmlessly but absorbed UV radiation is converted to radiation of a different wavelength and this could affect the plastic. In most instances, however, the time of exposure to UV is likely to be small and the effect on the plastic should be negligible.

UV screening agents can be either pigments or colourless organic compounds that absorb UV radiation. The most effective pigment is carbon black which gives optimum protection at concentrations of about 2–3%.

### A.2.9 UV stabilisers

UV stabilisers are added for the purpose of protecting the plastic itself and operate by interfering with the degradation chains initiated or catalysed by the radiation. The breakdown of a polymer when exposed to UV radiation involves a light-catalysed free radical mechanism which results in rupturing of the bonds along the polymer chain and degradation of the polymer's physical properties.

The possible effects of UV radiation on an unstabilised polymer include a reduction in tensile or impact strengths, dimensional changes, discolouration, surface cracking or crazing, powdering or an increase in electrical conductivity. These undesirable effects can be overcome by adding compounds which themselves absorb UV radiation and, at the same time, resist decomposition. The requirements of a material to be used for UV protection are as follows:

(a) the material must absorb UV light in the appropriate region
(b) it must be stable to UV light
(c) it must be stable under processing conditions
(d) it must be chemically inert
(e) it must be compatible with the polymer
(f) it should be neutral in colour
(g) it must be non-toxic (in food-contact uses)

A wide variety of compounds have been used as UV stabilisers including substituted acrylonitriles, benzotriazoles, substituted 2-hydroxybenzophenones and metallic complexes, such as nickel or cobalt salts of substituted phenols, thiobisphenol, phosphonates and dithiophosphinates. A more recent but important class of UV stabilisers is that of hindered amine light stabilisers (HALS) such as tetra-methyl piperidine, because they have received FDA clearance.

The UV stabiliser addition is normally incorporated at 0.1–0.5% of the polymer weight at the point of fabrication or during polymerisation.

# Closures

The subject of closures covers a very wide field and includes such container materials as paper, board, plastics, glass, metal and wood. However, we are concerned here with plastics closures for containers such as bottles and jars.

## B.1 Functions of a closure

Before looking at the various types of closure available, it might be helpful to consider some of the main functions of a closure.

1. The closure must provide an adequate seal until the contents are required for use. Usually this entails preventing escape of the contents and ingress of the external environment. The degree of seal tightness, however, is dependent on the product packed and many products may not need a completely hermetic seal.
2. In most instances it must be possible to open the container without difficulty and to reseal it properly when only part of the contents is used at a time. Alternatively, of course, the closure may be provided with a dispensing device, such as a spout, which is operated without removal of the closure. Sometimes, as with injectable fluids, the closure may be fitted with a pierceable wad.
3. The closure may need to provide a tamper-evident device to show whether it has been removed prior to use. This has become a very important area since the Tylenol incident that took place in the USA during September–October 1982. There were seven deaths in this incident, caused by capsules of Tylenol having their contents replaced, or partly replaced, by sodium cyanide. This has been followed by similar incidents in the UK. Apart from this, there is also the possibility that inspection of certain food packs on supermarket shelves out of curiosity, could destroy the sterility of the contents.
4. The closure may need to provide resistance to opening by children. In

this connection it is interesting to note that during tests, a plastic container fitted with a child-resistant closure, was opened by a very small child who bit right through the base of the container itself. This proves, if proof were needed, that there is no such thing as a child-*proof* container.

5. The closure must not affect the contents of the container, nor be affected by them. It should be resistant to any climatic conditions likely to be encountered and it may need to withstand conditioning or processing treatment, such as pasteurisation or sterilisation. The product/closure interaction is affected by the area of contact between the closure and the product and by the fact that many screw or snap-on closures may be fitted with an internal wad or liner so that the closure material does not contact the contents.

With narrow-necked bottles the area of contact is very small in relation to the volume of the contents. Since migration is proportional to area of contact, the resultant hazard is small. With wide-mouth jars, of course, the area of contact can be quite appreciable.

Where a liner, or wad, is used then the material of the main closure is practically irrelevant. What matters is the material actually in contact with the contents. The subject of wads will be returned to later.

6. Some closures may need to retain either a vacuum or an internal pressure.

7. The closure must often blend with the graphics of the container, enhancing the appearance and adding to the sales appeal. Most plastics can be self-coloured in a wide range of effects. The only common exception is phenol formaldehyde which can only be obtained in black, brown or dark colours such as maroon or blue. Polystyrene is particularly versatile and can be obtained in a wide range of metallic and pearlescent effects, in addition to normal colours.

Polypropylene has been widely used as a decorative closure because it is resilient enough to be able to have slight undercuts moulded in it. It is thus possible to mould undercut holes or channels into which can be snapped other plastics of contrasting colours.

8. The closure may also have to comply with certain performance requirements. Impact strength may often be a factor, for example, especially where conditions are severe, either on the filling line or in the distribution system. The closure may also have to be resistant to cracking and to creep in order to withstand excess torque during screwing-on (as with screw caps) or other internal forces. Thermosets, such as phenolformaldehyde or urea formaldehyde, are generally rather brittle but usually have excellent resistance to creep. HDPE has good impact resistance and is fairly rigid. Its creep resistance is limited, however, and this affects torque retention adversely.

Of the other plastics commonly used for closures, high-impact polystyrene has good impact resistance and creep resistance. Polypropylene homopolymer gives a good overall spread of properties provided that it is used at temperatures above 0 °C. Polypropylene copolymers are better in terms of low-temperature impact strength but have slightly lower tensile strengths than the homopolymer.

## B.2 Closure design

There are basically three main types of plastics closure, namely:

(a) screw caps
(b) snap-on caps
(c) plug closures

Of these, only screw caps can give a really positive gastight seal and even these are not in the same class as crimped-on or double-seamed metal closures. Snap-on and plug closures can be made liquid-proof but are recommended only for the less severe end uses.

### B.2.1 Screw caps

With screw caps, the underside provides the seal by being pressed against the top of the bottle neck or into its mouth. The pressure is generated by the screwing operation and must be evenly distributed and maintained to ensure a uniform seal round the whole of the edge of the cushioning material in contact with the rim. The higher the pressure (within reason), the more effective will be the seal but the pressure must obviously stop short of the point at which the cap can either break or deform, the bottle finish becomes chipped or the liner breaks down by splitting or collapse.

All standard screw closures can be applied automatically by a rotating head having a chuck that fits over, and grips, the loosely pre-applied cap and tightens it to the required torque. This torque is regulated by means of an adjustable friction clutch fitted to the head. Hand-fed single-head semi-automatic machines can apply caps at speeds of up to about 60 per minute whilst multi-head fully automatic rotary machines are capable of operating at speeds of 250 per minute or more. Caps can also be tightened by conveying the loosely capped bottles in a straight line between rollers or bands applied to their sides. Really fast speeds in excess of 400 caps per minute are possible with such a straight line capper.

The type of thread used depends upon the end use and on the moulding conditions but the one that has proved the most satisfactory is the buttress or saw-tooth thread. With this type of thread it is not possible to overscrew the cap or to strip the thread. The standard glass thread is not so satisfactory in use but is sometimes preferred by injection moulders because moulded caps with this type of thread can be pulled off the mould core. With the buttress thread, expensive unscrewing type moulds have to be used.

Types of screw threads can be further subdivided as follows.

### Screw caps fitted with wads

Wads consist of a resilient disc together with a facing material which is the sealing agent in contact with the bottle neck (and, of course, with the contents). Wads are still widely based on pulp board (and to some extent on composition cork) but these are being supplemented or replaced by expanded polyethylene.

Among the common wad materials are:

1. aluminium foil (usually laminated to a glazed imitation parchment paper – GIP)
2. PVDC (coated on to bleached kraft paper)
3. polyester film (laminated to a bleached kraft paper)
4. LDPE film (extruded on to bleached kraft paper)

It should be noted that metal facings can be lacquered if there is any risk of corrosion.

In the UK, two grades of expanded polyethylene are in use, with compression ratios that equate with pulpboard (10–20%) or composition cork (35–50%). In general, a low compression factor is recommended for use with narrow-necked containers and a high compression factor for wide-mouthed ones. This is because wide-mouthed containers are more likely to have a less even surface. Another factor is that composition cork develops a greater permanent set than does pulp board and this generally results in low cap-loosening torques after the bottles have been in stock for some time.

### Screw caps with flowed-in linings

Flowed-in linings may be in the form of annular sealing rings or overall lining discs. There are two main types of flowed-in compounds. One type is based on dispersions of PVC in plasticisers (plastisols); the other is based on colloidal dispersions of natural or synthetic rubber in water. The former are heated for a relatively short time at high temperature, while the latter require a longer heating time to drive off the water. Both types are available in different grades with a range of properties such as compression factor, heat resistance, product resistance and extractability. Flowed-in linings are now found in plastics caps as well as in metal ones.

Earlier problems associated with the retention in use of plastics caps (such as 'backing off') have largely been overcome by a variety of retaining features that prevent seal movement at the compression/interlock stage. As a result of such changes, plastics have made steady inroads into the ever-expanding field of carbonated drinks. Such closures often include a skirted tamper-evident feature which may be 'jumped-over' the bottle neck bead or formed under it by a heating/cooling action.

*Wadless screw caps*

As mentioned in Chapter 2, polypropylene is very resilient. This means that the material will 'give' sufficiently to take up inequalities in the neck sealing surface but is resilient enough to push back strongly and maintain a seal. Even with polypropylene's resiliency, the cap sealing surfaces have to be fairly thin sectioned in order to achieve the correct flexibility. Two possible designs of wadless closure are shown in Fig. B.1.

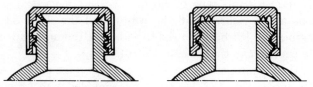

**Figure B.1**    Examples of polypropylene linerless closures.

### B.2.2 Snap-on caps

Snap-on caps are made from thermoplastics because flexibility and resiliency are the predominant requirements. They are normally used where a quick and easy seal or reseal is wanted. The cap is forced over a cooperating bead on the mouth of the container so that the cap hugs the bead tightly and uses the whole surface of the bead as a sealing area. Where liquid-tightness is required, the tolerances of the container neck and the cap have to be more tightly controlled than when using screw caps. Where automatic closing equipment is used, the basic containers will have to be of sufficient strength to withstand the top-loading pressures involved.

### B.2.3 Plug closures

In the plug closure type of seal an interference fit stopper is inserted into the mouth of the container. It is the oldest method of sealing a bottle. Natural corks – used for hundreds of years – are still being used today.

The modern day counterparts are, like snap-on closures, made from thermoplastics because of the elasticity required. They provide a seal by being made oversize and hence they undergo compression when forced into the bottle neck. The outside surface of the plug is often moulded to give fins or serrations thus giving even greater flexibility and allowing a greater degree of interference fit to be achieved. In addition, each serration or fin acts as a separate seal.

## B.3 Special closures

The need that often arises for tamper-evident closures was mentioned earlier. There are many designs possible but one general type utilises a press-on cap with a deeper than usual skirt, the bottom of which snaps over a circumferential bead on the bottle neck. Above this bead, the cap skirt is scored and perforated in such a way that a narrow strip can be torn off round the cap, leaving a short skirted cap above the bead and a ring of plastic below as evidence of opening. LDPE is the material most commonly used for this type of closure because of its flexibility. If the perforations are not carried all the way round the skirt, then a small connecting piece is left between the cap and the retained ring and hence provides a captive cap that cannot be lost.

Another type of tamper-evident closure relies on the use of a more rigid plastic, often polypropylene. A screw cap is moulded with a ratcheted ring attached to its skirt by thin retaining spokes. The neck of the bottle is similarly ratcheted in such a way that the cap can be screwed down but not up again without breaking the retaining spokes. Once again, a ring of plastic is left behind and it is obvious that the container has been opened.

Another novel type of closure is the child-resistant closure (CRC), also mentioned earlier. They are normally closures that require two coordinated actions for removal and they usually fall into one of the following basic types.

1. *Press and turn.* Here the cap must be pressed against the bottle and rotated at the same time. The Clic-Loc closure, based on this mechanism, has an additional feature in that it gives an audible alarm when anyone attempts to open it by normal turning action only.
2. *Squeeze and turn.* This is a double cap with a free-rotating outer which must be squeezed to engage the inner cap when the closure is to be unscrewed.
3. *Combination lock.* In this type, two parts of the cap must be lined up according to the marks shown on the cap.
4. *Restraining ring.* This is a two-piece cap in which a top piece and a

restraining ring are screwed together. The combination then rotates freely on the bottle. Since the restraining ring is permanently attached to the bottle, opening can only be achieved by holding the restraining ring in one hand and unscrewing the top from it in the other.

5. *Press and lift.* In this type a downward pressure on the cap relaxes the grip of the cap's edge on the bottle and it can then be lifted free.

## B.4 Dispensers

An extremely wide range of spouts, pourers and patented dispensers of all kinds is available. Thermoplastics are normally used, because of the design versatility offered by thermoplastic moulding processes, particularly injection moulding. The most common material is LDPE because of its low cost, flexibility and high impact strength. In food packaging uses it should be noted that contact with the food may only be intermittent as the dispenser is out of contact except during the act of pouring.

In other applications, parts of the dispenser may be in contact all the time. A common situation is where the dispenser is fitted to the top of the container and is initially in contact with the contents all the time. Once some of the contents have been used, however, contact becomes intermittent and occurs only during the act of pouring. Two of the very many different types of dispenser/closure available are shown in Fig. B.2.

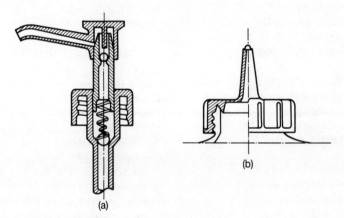

**Figure B.2** (a) Pump closure and (b) snip-top cap.

# INDEX